PRACTICAL BUILDING CONSERVATION
VOLUME 5
WOOD, GLASS AND RESINS
AND
TECHNICAL BIBLIOGRAPHY

Practical Building Conservation Series:

PRACTICAL BUILDING CONSERVATION

English Heritage Technical Handbook

VOLUME 5

WOOD, GLASS AND RESINS
and
Technical Bibliography

John Ashurst
Nicola Ashurst
with
Patrick Faulkner
Hugh Harrison
Jill Kerr
and
Edmund King

Graphics by Iain McCaig

Gower Technical Press

Published by
Gower Technical Press Ltd,
Gower House,
Croft Road,
Aldershot,
Hants GU11 3HR,
England

Reprinted 1989

British Library Cataloguing in Publication Data

Ashurst, John
    Practical building conservation: English
    Heritage technical handbook.
    Vol. 5, Wood, glass and resins
    1. Great Britain. Building of historical
    importance. Conservation. Techniques
    I. Title   II. Ashurst, Nicola
    720'.28'8

    ISBN 0 291 39776 X

Printed and bound in Great Britain at
The Camelot Press Ltd, Southampton

CONTENTS

FOREWORD

by Peter Rumble CB, Chief Executive, English Heritage

Over many years the staff of the Research, Technical and Advisory Service of English Heritage have built up expertise in the theory and practice of conserving buildings and the materials used in buildings. Their knowledge and advice have been given mainly in respect of individual buildings or particular materials. The time has come to bring that advice together in order to make available practical information on the essential business of conserving buildings — and doing so properly. The advice relates to most materials and techniques used in traditional building construction as well as methods of repairing, preserving and maintaining our historic buildings with a minimum loss of original fabric.

Although the five volumes which are being published are not intended as specifications for remedial work, we hope that they will be used widely by those who write, read or use such specifications. We expect to revise and enlarge upon some of the information in subsequent editions as well as introducing new subjects. Although our concern is with the past, we are keenly aware that building conservation is a modern and advancing science to which we intend, with our colleagues at home and abroad, to continue to contribute.

The Practical Building Conservation Series

The contents of the five volumes reflect the principal requests for information which are made to the Research, Technical and Advisory Services of English Heritage (RTAS) in London.

RTAS does not work in isolation; it has regular contact with colleagues in Europe, the Americas and Australia, primarily through ICOMOS, ICCROM, and APT. Much of the information is of direct interest to building conservation practitioners in these continents as well as their British counterparts.

English Heritage

English Heritage, the Historic Buildings and Monuments Commission for England came into existence on 1st April 1984, set up by the Government but independent of it. Its duties cover the whole of England and relate to ancient monuments, historic buildings, conservation areas, historic gardens and archaeology. The Commission consists of a Chairman and up to sixteen other members. Commissioners are appointed by the Secretary of State for the Environment and are chosen for their very wide range of relevant experience and expertise. The Commission is assisted in its works by committees of people with reputation, knowledge and experience in different spheres. Two of the most important committees relate to ancient monuments and to historic buildings respectively. These committees carry on the traditions of the Ancient Monuments Board and the Historic Buildings Council, two bodies whose work has gained them national and international reputations. Other advisory committees assist on matters such as historic gardens, education, interpretation, publication, marketing and trading and provide independent expert advice.

The Commission has a staff of over 1,000, most of whom had been serving in the Department of the Environment. They include archaeologists, architects, artists, conservators, craftsmen, draughtsmen, engineers, historians and scientists.

In short, the Commission is a body of highly skilled and dedicated people who are concerned with protecting and preserving the architectural and archaeological heritage of England, making it better known, more informative and more enjoyable to the public.

ACKNOWLEDGEMENTS

The authors gratefully acknowledge the assistance of Dr Clifford Price, Head of the Ancient Monuments Laboratory, English Heritage, in the reading of the texts.

1 STRUCTURAL AND DECORATIVE WOOD IN BUILDING

Included in this chapter is a part reprint of *Timberwork*, published in 1965 by HMSO for the Ministry of Public Building and Works and written by **Patrick Faulkner**, lately Superintending Architect, Ancient Monuments. The original text has been extended and amplified by John Ashurst. RTAS is grateful to **Hugh Harrison** for the contributions on the conservation of decorative wood surfaces.

1.1 WOOD IN BUILDING

In Britain the wood generally selected for structural and non-structural work in buildings was oak, although in some locations and at some periods sweet chestnut has been used in the same ways. Many other hardwoods have long been available, but the early abundance and durability of oak made it the natural choice. In addition to native oak, further supplies were obtained from Europe. From the sixteenth century onwards sofwoods, principally fir and pine, were also imported. In the nineteenth century softwood began to be imported in large quantity from North America.

Unseasoned, 'green' oak was commonly used for building and is still used, and often recommended, for repairs to historic timber-framed buildings. Seasoned oak is unlikely to be available, in any case, in sections over 100 mm (4″) square. When seasoning was essential for work such as joinery, oak, or other timber, it would be stripped of bark and submerged in running water for a year or more. During this time the acids and saps could be replaced by water, so that when the wood was exposed again to the air, shrinkage and distortion were reduced to a minimum.

Wood is formed in the growing process of a tree. Trees have been described, picturesquely, as 'food factories', whereby moisture and carbon dioxide are converted by the sun's energy to cellulose and lignin. High magnification of a section of wood shows that these substances are laid down in cellular form, with perforated cell walls permitting the free passage of moisture. Every year the girth of a tree is increased as successive cell layers are added to the external perimeter growth rings. The growth is intermittent, taking place rapidly in the spring ('early-wood') and more slowly in the summer ('latewood'), distinguished usually by light and dark wood. These growth rings record the seasonal variations and rate of growth. In general, rapid growth is associated with wide growth rings , loss of density and loss of strength. Close growth rings are therefore desirable. BS 1186; Pt 1, 1971 requires at least 8 growth rings per 25 mm. The core of the tree is referred to as heartwood and the outer layer, lighter in colour, are referred to as sapwood. This sapwood, 25 mm to 150 mm in thickness and containing starch, sugars and water, is particularly attractive to wood-rotting fungi and insects, although heartwood can also be attacked.

The objective of timber treatments must be to reduce the attractiveness of wood as a food source and to overcome its vulnerability to natural enemies.

Much excellent literature is readily available on wood-destroying fungi and insects. Some of the references are given at the end of this chapter and in the bibliography (Chapter 4). This chapter will, therefore, make summary references only to the main problems and methods of treatment.

1.2   WOOD-ROTTING FUNGI

Fungi obtain their organic food from the dead or living parts of other plants or animals. Fundamental for germination in a suitable food source are a moisture content over 20 per cent, the presence of air and a suitable temperature. The optimum temperature for most wood-destroying fungi is between 20° and 22°C (68° and 72°F).

The most obvious part of a fungus is the reproductive fruit body, which may appear in several forms, but often as flat plates or brackets. The presence of a fruit body (the 'sporophore') on wood indicates that an elaborate system of tubes or threads (the 'hyphae') has already been established. A mass of these threads (a 'mycelium') is clearly visible and is much more commonly seen than the fruit body. The development of the fruit body may indicate a crisis in the development of the fungal growth; from its spore-bearing surface (the 'hymenium') millions of microscopically small seeds are dispersed, only one or two of which are required to infect new wood. Germination will take place if the conditions described above are present. The hyphae branch out from the seed, thinning the cell walls and eating holes through them, producing, as they spread, enzymes which break down wood components into soluble materials such as glucose. The obvious effects of this breakdown are colour and weight change, splitting along the grain and cross-cracking ('cuboidal cracking'). Two of the most common wood-rotting fungi found in buildings are the 'cellar fungus' (*Coniophora puteana*) and true 'dry rot' (*Serpula lacrymans*).

Serpula lacrymans requires quite specific conditions in which to flourish, but it has the ability to modify its immediate environment, when necessary, by conducting nutrient and moisture through its strands. These strands, which can travel

---

The cellar fungus *(Coniophora puteana)* (a 'Wet Rot')

Location:	Very damp buildings. Often in timber in wet masonry or in association with wet ground. Outbuildings and fences are typically affected.
Victim woods:	Softwoods and hardwoods.
Visual evidence:	Blackish-brown strands spreading fan-fashion over the suface of timber and walls. Fruiting bodies rarely seen.
Effect on wood:	Dark brown colour, cracking along grain, cross cracking.

True dry rot *(Serpula lacrymans)*

Location:	Only in buildings (or wooden ships).
Victim woods:	Usually softwoods, but sometimes hardwoods.
Visual evidence:	Spectacular fruit bodies in the form of flat plates with red hymenium and white edges. White masses of hyphae, when the growth is very active, which may be covered with 'tears' of moisture, hence 'lacrymans'. Matted grey skins and thin, grey strands which may be under as well as across plaster. Large strands conducting food and water. ('rhizomorphs') may be up to 6 mm in diameter.
Effect on wood:	Light brown colour, dry and embrittled. Deeper cuboidal cracking than wet rots.

---

through and over a variety of materials, make serpula lacrymans a persistent and potentially much more hazardous fungus than others. Even when isolated from wood they may lie dormant for several years in suitable conditions and may infect installed wood which becomes wet. Development of the fungus and destruction of victim wood are quite rapid.

Standard treatment procedures for infected wood (rot)

Standard remedial treatment may be summarized as follows:

1   Expose all timber to determine the full extent and type of attack. In the case of dry rot, check behind plaster on both sides of a wall and over ceilings.
2   Identify and locate sources of moisture and carry out the necessary remedial work to joints, flashings, soil levels, drainage etc.
3   Cut out all infected timber. In the case of dry rot cut away timber within 0.5 m (20″), even if it is apparently sound, and burn the wasted material.
4   Thoroughly remove debris such as dust, shavings and insulation which may be infected.
5   Provide adequate ventilation of suspended floors, eaves and other roof spaces.
6   In the case of dry rot, remove plaster containing or covering fungal strands to at least 600 mm (24″) beyond the last visual evidence and brush clean or

3

Immersion of large section oak beam in a specially constructed dip tank filled with organic solvent preservative. (Photograph: DAMHB Archive)

vacuum the wall surfaces. Irrigate the wall with a suitable fungicide by drilling boreholes at 250 mm (10″) centres down into the wall above and around the infected area to within 50 mm (2″) of the opposite wall face. Walls over 250 mm (10″) thick are to be bored and flooded from both sides. In severe cases the entire area, not just the periphery, is to be flooded.

7 Flood sound timber in the vicinity of all rot by three brush or coarse spray applications, first removing any paint or varnish from the surfaces. Treatment rate with organic solvent preservative should be at the rate of 1.0 litre per m^2 of flooring and 2.0 litres per m^2 of roofing.

8 Ensure that new wood is well-seasoned and has been treated by ten minutes of immersion in preservative, or ten minutes' immersion of end grain and three floodcoats. Each spray or brush flood coat must be applied before the subsequent one has dried.

9   Treat selected specified areas with fungicidal paste at the rate of 0.5 to 1.0 kg per 5 m^2 surface area.

Environmental control in historic buildings

The impact of this kind of treatment, especially of dry rot, on a historic building which has been infected can be considerable both in terms of destruction and, in the case of irrigation, of built-in problems associated with drying out and salt migration. It follows that, although the standard procedures must be followed in some respects, especially on inspection and removal of actual decay and treatment of old and new timber, some compromise is often desirable. In some cases this compromise has been acheived by 'environmental control' and monitoring.

The role of water in creating conditions favourable to fungal growth and insect attack is well established and well known. It is not surprising, therefore, that much emphasis must always be laid on the need to keep moisture contents in buildings within 'safe limits' if the wood is to be protected. Wood will not decay as a result of fungal attack if it retains a moisture content below 20 per cent. Typical moisture contents of timber elements in a domestic property may seasonally range from 12 per cent to 18 per cent on the ground floor, 10 per cent to 15 per cent on the upper floor, and 12 per cent to 25 per cent in the roof space, staying largely within safety limits. Locally, however, timber such as wall plates and beam or rafter ends, or fixings in a cellar, may be well above these levels and almost certain candidates for infection. Timber treatments and timber replacements in such situations are a waste of money unless the source of water is cut off. When this has been acheived by remedial work on roof coverings, rain and soil water disposal systems, flashings, joints or damp course insertion, it should be possible to think in terms of minimum treatment and, most significantly, minimum replacement of original wood. Much 'overkill' in remedial treatments, especially in dry rot cases, still continues because of the persistence and tenacity of the fungus when well established and the 'belt and braces' practice inevitably linked to treatments which carry a guarantee. Control of environmental conditions as a means of containing potential fungal spread, rather than wholescale destruction and sterilization, is one of the most important developments of the past few years. When moisture has been denied access, as far as possible, to a former victim building, observation of the vulnerable zones can continue by built-in moisture probes which can be read individually or, in a large building, be linked to a central monitoring system. Hidden areas susceptible to fungal growth can be provided with permanent spy-holes through which fibre-optic inspections can be made. This kind of controlled approach to buildings where dry rot has occurred will do much to prevent needless destruction of old timber and plaster and, if properly maintained, is a far more positive guarantee against further trouble than the more traditional ones.

The danger of the approach lies in the built-in human factor. Unless inspections are regularly and conscientiously carried out and the results logged and heeded, the potential for loss is far greater than would have been incurred by standard remedial treatment. Research still continues into methods of control.

1.3 WOOD-DESTROYING INSECTS

Although, on a worldwide basis, termites cause the greatest amount of damage to timber in buildings, attack on timber in the UK and in most temperate regions is by insects, usually beetles, the Coleoptera. The most troublesome are the beetles which inhabit timber in their grub stage. During this period, after the grub is hatched, much damage is caused as the grub burrows its way into the wood, eating and growing for periods ranging from two to ten years. Digested wood is excreted in the form of pellets and dust. Towards the end of this destructive phase the grub enters a pupal stage, lasting a few weeks, before emerging, usually in the spring and summer, as an adult beetle. The beetle's emergence from the wood leaves the characteristic flight hole in the surface of the wood. In historic timber the most common wood-boring insects are the death-watch beetle and the furniture beetle. Of the other common types, brief mention may be made of the very destructive house longhorn beetle, *Hylotrupes bajulus*, found in the sapwood of some softwoods, but geographically limited in the UK to some parts of the south of England; the lyctus (powder post) beetle, found in the sapwood of some hardwoods; and the weevils *Pentarthrum huttoni* and *Euophryum confine*, found in damp and decayed hardwoods and softwoods.

The critical identification data of the two major pests may be summarized as follows:

Death watch beetle — *Xestobium rufovillosum*
Wood attacked: Hardwoods, or softwoods adjacent to hardwoods, especially in damp, unheated buildings, since a certain amount of moisture and fungal decay are necessary for the grubs to be able to live on the wood.
Grub: White, 4 – 6 mm long. Remains in the wood four to five years, or even as long as ten years.
Pupal stage: Grubs change into the pupal stage in the autumn and then, after two to four weeks, into adult beetles. The beetles remain in the pupal chamber until the spring.
Adult beetle: The beetle bores its exit 'flight' hole in the spring, although emergence may continue into September. The beetles are 6 – 8 mm long and the flight hole typically 3 mm.
Bore dust: Bun-shaped pellets.
Eggs: Mating takes place very soon after emergence. The female lays 40 – 70 eggs, which hatch in 2 to 8 weeks.

Furniture beetle – *Anobium punctatum*
Wood attacked: Hardwoods and softwoods, including plywoods and compositions containing natural glues.
Grub: White, 2 – 4 mm long. Remains in the wood two years.
Pupal stage: Grubs change into the pupal stage as spring approaches. After a period of two to three weeks in the pupal chamber hollowed out by the grub, the adult beetle is hatched.
Adult beetle: The beetles emerge through their exit 'flight' holes from May to September. The beetles are 2 – 6 mm long and the flight hole 2 mm.

Smoke generators thirty seconds after ignition, showing smoke columns reaching the roof spaces. At this stage, contact insecticide is being deposited on the downward facing timber surfaces. In another fifteen seconds the smoke will begin to return towards the floor, distributing insecticide on the upward facing surfaces. This treatment is required twice a year during the flight season of the death-watch beetle for up to ten years to achieve an effective control. (Photograph: DAMHB Archive)

Standard treatment procedures for infested wood (insects)

Standard remedial treatment may be summarized as follows:

1   Expose all timber possible to determine the full extent and type of attack. In the case of attacked floors enclosed by plaster and boarding, remove boards at 0.5 m spacings and perimeter boards and skirtings.
2   Vacuum clean away all dust and debris.
3   Cut away severely decayed wood and replace with sound, pre-treated wood.
4   Treat accessible timbers by applying insecticidal fluid to BS 5707 to saturation at 1.0 litre per m^2 of flooring and 2.0 litres per m^2 of roofing.
5   Treat both sides of floorboards by brush or spray before replacing.

Smoke deposited insecticides

For many years now, contact insecticides have been deposited on inaccessible (or difficult of access) wood surfaces in buildings using smoke as a carrier. This system has shown itself to be effective in the control of death-watch beetle populations when treatments have been carried out on an annual basis.

Smoke generators have generally been used to deposit stomach poisons such as Lindane, but growing concern and legislation on the effect of such poisons on bats and birds in roof spaces and towers have led to the increasing used of the synthetic pyrethroid Permethrin instead.

Most smoke deposited treatments have been carried out on open roof frames. The building should be as effectively sealed as possible against smoke losses, although in any building allowance must be made for unavoidable losses in calculating the amount of insecticide required. Suppliers will recommend the appropriate number and size of smoke generators for the volume of building or roof space to be treated.

The treatment should be carried out, if possible, on a windless day when the temperature is below 15°C (60°F). When it is ignited, a smoke column rises from the generator. During the upward movement, insecticide is deposited on the downward facing timber; as the smoke reaches the roof covering it cools and begins to descend, this time depositing on the upward facing surfaces.

The building should remain sealed for at least twelve hours (overnight is a practical period). When it is opened up, an industrial vacuum may be used to remove the insecticide from the floor, the operative wearing a gauze mask against dust. Dust sheets should be used to cover soft furnishings and other articles in the room which are not easily vacuum cleaned.

The treatment should be timed to be completed by April, when the beetles emerge. If possible, a second treatment should be carried out in May, to coincide with the period of mating and egg laying. The procedure must be carried out for at least seven consecutive years, and preferably ten, to take account of the life cycle of the beetle.

An alternative method of depositing Permethrin which may be used in smoke-sensitive areas is high pressure water, delivered as a mist.

1.4   TIMBER PRESERVATIVES

Preservatives fall into three categories, as follows:

Category	Example
TO: tar oil	Creosote
OS: organic solvent	Pentachlorophenol Napthenates Tributyl tin oxide Gamma-hexachloro-cyclohexane (Lindane) Synthetic pyrethroids (e.g. Permethrin) Organo boron compounds
WB: waterborne	Copper chrome arsenic

Timber preservatives can be applied in a variety of ways:

- By brushing or spraying. Slow evaporation and greater penetration result from using higher boiling point solvents, but there is always a flammability risk.
- By the additional use of bodied mayonnaise emulsions on heavy section timbers. A toxic reservoir is maintained under a hard skin which forms on the paste.
- By the insertion of borate rods, especially in external joinery. The rods are inserted into sealed drillings. The preservative diffuses when the wood becomes wet.
- By injecting into plastic nipples incorporating a non-return valve. The nipples are set into 10 mm (½″) holes down into the wood.

1.5 CAUSES AND EFFECTS OF FAULTS RELATED TO CONSTRUCTION

The historical value of timber

The preservation of timber in a historic building presents particular problems of both technique and approach. The historic value of the building may, indeed, be dependent on the extent and manner in which it utilizes timber in its construction and decoration. To this extent, if such a building is to maintain its historic interest, every effort must be made to preserve surviving timber in its original context, or, if this is structurally impossible, to replace or restore in the manner in which evidence shows the timber to have been first employed. To do less is materially to impair the value of the building. Preservation and repair are two different aspects. Preservation implies taking steps to ensure that timber (old and new) will be protected from future decay, an operation that becomes increasingly important as the intrinsic value of the individual timber rises. It must be stressed that this value is not necessarily dependent on age or elaboration; it may lie solely in the manner of its use. Repair in this context may be defined as the repair of the timber in such a manner that it may perform once again the function for which it was originally designed. Here the function is taken primarily to be structural; it may also be aesthetic.

Framcd buildings

In any building there are certain danger spots where faults may develop as a result of the decay or failure of incorporated timber. Most of these are well known, but in old buildings there may be some not so widey appreciated. Those which occur in wholly timber buildings are generally self-evident but, even here, are apt to be overlooked. The symptoms are generally the distortion of the frame or the cracking and displacement of panel fillings in both internal and external walls, the cause of which may be traced to broken or, more often, cut members. Common examples are the cutting of tie beams and braces of roof trusses in order to insert doorways between attics; the removal of a main cross beam to insert a staircase; the removal of cross frames and their braces to open up two rooms into one and, in fact, any one of the usual 'improvements' to which a building is subject and which result in the severance of a structural member of the frame. Even if there has been no such removal of parts, the decay of those hidden can prove equally harmful. Cracked plaster (or distorted panelling) may be the clue to the position of a buried brace in a cross frame, the tenons of which have decayed and pulled, dangerously weakening the structure. Some close-studded buildings have braces halved over the backs of the studs hidden by the exterior panel filling and by internal wall plaster. These, too, should be sought out as possible origins of structural failure.

Roofs over brick or stone buildings

By far the greatest number of failures in timber roofs arise through decay of the timbers resulting from failure of the roof covering. Inspection of the roof itself will reveal these. Less obviously, the real fault may lie in the supporting walls rather than the roof. Some roofs were designed to exert an outward thrust on the walls, others were so framed as to give a vertical resultant thrust. In either case, deteriorated, undermined or weakened walls may, by movement, throw unreasonable loads on the truss members. Pulled tenons and buckled and bent struts are as likely to be caused by external circumstances such as these as they are by inadequate design of the truss. The answer may lie in reinforcing the walls rather than the timber truss, which may only need repair.

Built-in timbers

For many of several reasons an ashlar or brick facing is often built onto a rubble masonry or brick backing with little or no effective bond between the two. The facing is provided with its own lintels or rubbed brick arches, whilst the backing may be carried on hidden timber lintels across the opening. The decay of these hidden timbers, whether they be bond timbers or lintels, allows the backing to sink and may throw a heavy load on the facing, which will then either bulge or fracture. Such a bulge or fracture is often the clue to decayed timber within the wall; the mere tying back of the bulge will not be a satisfactory repair.

Trussed partitions and trussed beams

The builders of the eighteenth and nineteenth centuries made much use of trussed partitions and many ingenious forms of trussed beams across the wide spans of ground and first floor state rooms. Later alterations sometimes paid little heed to the structural importance of these, and doorways were cut through the former and the latter were pierced for chandelier chains and the like. The result may be seen in sagging ceilings or, if the partitions are panelled, in split panels. Such changes should always be looked for in all cases where there is evidence of some structural failure and especially in an area which includes smaller rooms over larger.

1.6 REPAIR OF FRAMED BUILDINGS AND ROOF TRUSSES

The frame as an entity

The repair of a framed structure, whether it be a wholly framed building or a roof truss, will always call for an appreciation and understanding of the frame as a whole. The failure of any one member of the frame is liable to throw undue stress on the remainder — hence the importance of discovering why a timber has failed before embarking on its repair. It is, in general, reasonable to assume that the original design of the frame or truss was adequate. It follows that, if each individual member can be so repaired as to be capable of playing the part for which it was intended, the frame again becomes a stable entity, structurally sound, without the assistance of extraneous reinforcement in the shape of ties, posts, struts etc., in steel or any other alien material. The acceptance of this principle and its application, following a careful appreciation of the structural implications, will often enable less extensive repairs to be carried out than might at first sight be thought desirable, not only saving expense but preserving to a greater extent the historical validity of the building.

Members of the frame or truss

The essential objective of a repair is to restore the structural strength of the member and of the joints which connect it to the frame of which it is a part. First the timber itself: many timbers, particularly in medieval buildings, have what would be considered an excessive safety factor today. It is possible for them to have suffered quite considerable decay or damage and yet still retain sufficient strength to play their part in the frame, provided that this still forms an entity. It follows that, if they can be preserved as they stand, there is no need for repair. This requires some consideration of the part they are playing and where they are damaged; a beam, for instance, may be decayed at the sides or be cut into, but have sufficient depth still remaining to permit its taking the designed load. On the other hand, a pipe chase across the underside of an otherwise well preserved beam may so reduce the effective depth as to render it dangerous. The repair, therefore, of an individual member must take into account the part it plays within the frame and the stresses it is called upon to withstand.

Repair of compression members

In the repair of a compression member it may be necessary to cut out a decayed section and insert new wood, in order to build up the necessary cross-sectional area, in which case care must be taken to ensure that the cuts are made in a direction normal to the grain and that the lie of the grain in the inserted portion corresponds to that on either side. An accurate fit for the insertion is important, as the joint is load-bearing, and complete contact must be made. This may be easier if a stepped joint is used. The joint should be grouted up with water-proof and boil-proof glue. Pins may be used to locate the insertion but should be not be relied upon to transmit any load.

Repair of tension members

In the repair of a tension member a different approach is necessary, though again the criterion must be the adequacy of the surviving cross-section area. Timber has a suprisingly high tensile strength which depends on the continuity of its grain structure. A wide shake or split along the grain, for instance, will not materially reduce the tensile strength of the timber, whereas a cut or knot or drilled hole will

Typical joints used in carpentry repairs

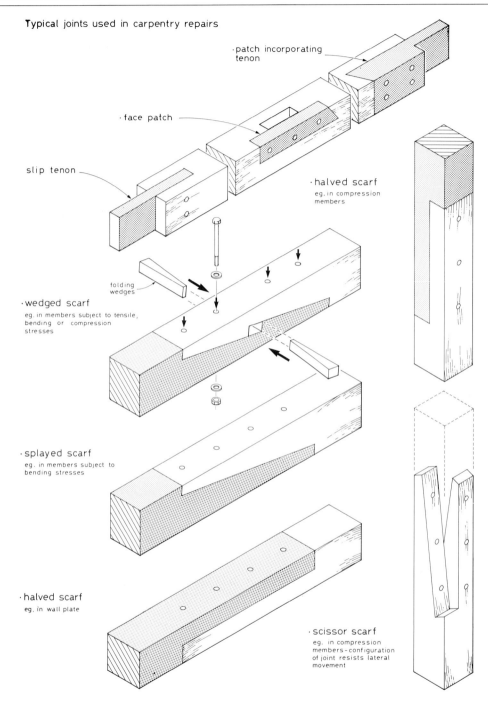

·patch incorporating
tenon

·face patch

slip tenon

·halved scarf
eg. in compression
members

folding
wedges

·wedged scarf
eg. in members subject to tensile,
bending or compression
stresses

·splayed scarf
eg. in members subject to
bending stresses

·halved scarf
eg. in wall plate

·scissor scarf
eg. in compression
members – configuration
of joint resists lateral
movement

Figure 1.1   Scarf joints in traditional timber-to-timber carpentry repairs
*In the situations exampled, these types of repairs should always be considered first,
rather than resorting to substitute materials.*

do so. The most satisfactory joint for a purely timber repair, if the memeber is not capable of taking the stress, is to cut out the damaged portion entirely and insert a new section joined by means of a mortice and tenon at each end. The use of steel reinforcement is considered below.

Repair of beams
A member subject to bending stress, for example a beam or rafter, must be repaired in such a way as to be able to withstand this. Most often damage occurs at the ends and the repair may be effected by a new portion scarfed to the old. If it is to be repaired at its centre a halving or plating (in timber) is more appropriate, the joints being vertical to maintain the essential depth. Fixings, again, would be by pins well grouted with glue.

Repair of joints
Failure at the joints between members is as dangerous to a frame as the failure of the members themselves; they are, moreover, points particularly vulnerable to fungal and insect attack. In the repair of joints the same basic principles must be followed: they must be repaired in such a manner as to enable them to carry out the functions for which they were designed. This usually implies reinstatement. If the timber is so decayed or damaged at this point that renewal is in any case essential, then a straightforward copy of the original joint must be made, pinned in the same manner. If, however, only part of the joint has suffered then it may be repaired. The tongue of a tenon, for instance, may be cut off, a mortice made in the sound end of the timber and a false tenon inserted, pinned at both ends. If this is to be done *in situ* then the new mortice must be made sufficiently deep to allow the false tenon to slide clear of the old mortice and then be wedged in position when located. The false mortice may be grouted up with glue to eliminate unnecessary voids, but the real joint should still permit the structural movement inherent in a timber frame. Similarly, if the walls of the mortice have been damaged they may be built up with insertions. The repair of the dovetail joint between tie beam and wall plate may be treated in the same way by applying the principle of reinstatement. Building up the shoulders of the dovetail, either male or female, will restore the joint to its original strength and purpose. Water- and boil-proof glue should always be used.

1.7   DISTORTION OF TIMBER STRUCTURES

Framed buildings
A framed building may have acquired a very considerable distortion through the set of its individual timbers. Medieval buildings, particularly, were often built of green timber, which may twist quite alarmingly. In fact there is no harm in this as long as the joints remain sound. Nevertheless, distortion may be a sign of failure of some part of the frame and this, at any rate, should be suspected until it can be proved to be unfounded. No attempt should be made to correct distortion of long standing unless it is possible to free the entire frame (or truss) from fixed or

inflexible structures, in which case careful jacking up may serve to restore the frame to its original position. If site conditions permit, dismemberment of the frame and re-erection may often prove to be the more economical course. If distortion can be shown to have been caused by the mutilation or failure of a member this must, of course, be corrected first.

Beams

The distortion of a beam may well be due again to initial set and as such needs no repair. If the floor it carries is so uneven as to be unusable this can only be firred up; the beam itself should be left undisturbed. Distorted beams carrying decorative ceilings, either of plaster or wood, should generally be repaired to the shape they have assumed, because if the distortion is of long standing a plaster or wood ceiling will have followed the set and will only revert to a true line with the risk of considerable damage.

Thrust roofs

Many medieval and some later buildings have roofs without tie beams which exert an outward pressure on the walls. Spreading walls may so distort the roof structure as to fracture or damage the timbers. The repair of the truss and its members is treated as the repair of any other framed structure, but before any such repair is made the cause of the failure must be removed. Generally this is because the walls themselves have become too weak to withstand the side thrust. The cure is not necessarily to introduce ties but to strengthen the wall, by grouting if they are thick enough but weak, or perhaps by the introduction of a reinforced concrete plate designed as a horizontal beam tied at cross walls to replace the original timber plate.

1.8 BUILT-IN TIMBER

Non-structural timbers

Timbers built into or enclosed by brickwork or masonry are the source of much trouble in old buildings. As with other timbers, repair must only be undertaken when their purpose is fully understood. Some may be described as dead or non-structural timbers such as fixing blocks and isolated pads for beams. All these may simply be replaced by some other suitable inert material that will fulfil the same purpose: there is probably little virtue in their retention.

Structural timbers

A second class of built-in timbers is comprised of those which are live and form an essential part of the structure. Their repair requires detailed consideration. From the earliest times to the nineteenth century timber has been used in brick and masonry buildings as a reinforcing material, especially in high structures where, at each stage, the rigid floor frame may be used as a horizontal stiffener with its peripheral member built in as the work proceeds. This technique was of particular value to the Jacobean, Stuart and Early Georgian builders, who were thus able to introduce tall windows, using the horizontal members of the floor frame as lintels and reducing the brickwork or masonry to be mere piers between windows, which, without the support of the floor frame, would be so slender as to be in danger of buckling. Provided the timber is in good condition and the continuity of the floor frame is not broken by damage or decay, this is a perfectly sound method of construction. Repair, therefore, must aim at restoring the purpose of the horizontal frame. Much elaborate reinforcement of the wall can be avoided if this is recognized and the continuity of the peripheral member is maintained. Site conditions may make it difficult to replace timber with timber (though this should never be ruled out automatically), and steel or resin may be more practical materials for introduction in sections. The joints between sections must be continuous and the cross beams properly tied in to achieve a full frame. The same consideration must be given to other buried structural timbers. Bond timbers, as their name implies, are there to bond a loosely constructed wall together and their replacement must continue to do so. Dubbing up the chase or building in a course of brickwork is inadequate, and some reinforcement must be provided. Relieving lintels, with the lower plate acting as a spreader beneath a non-framed roof, are other examples of structural timbers which may be wholly buried and whose repair calls for similar treatment and understanding.

1.9 TREATMENT OF EXTERNAL TIMBER

Timber-framed buildings

The principal danger points in an exposed timber frame are the ledges formed by the horizontals, the undersides of the sole plates and the junctions between panel fillings and studs. Where new (or secondhand) timber is being introduced the horizontal surfaces may be given a slight weathering provided the grain is sufficiently close; otherwise, by far the best protection is a Code 4 lead weathering taken up under the rendering (if this exists or, if there is nogging, in the first brick joint). A renewed sole plate should always be bedded in lead in preference to bitumastic felt, and where possible lead should be inserted under an old plate. For the junction between fillings and studs a flexible mastic is best, if possible against a slightly recessed batten fixed to the side of the structural timber. On no account should the exposed surface of the timber be sealed. Bitumen or paint coverings tend to direct water to and hold it in the joints, thus providing a potential source of decay. A clear preservative may be applied provided it has no sealing properties. The filling of shakes with slips can be actively harmful by holding water and providing, in the void behind, a breeding ground for beetles. Only where the shake is so formed as to lead surface water into the interior of the timber should it be closed with mastic or a filled resin.

External timber features

Timber used externally for decoration and cladding, such as in cornices, door surrounds and hoods and weatherboarding, was invariably painted and after analysis to determine the original type and colour should be painted again, but

15

ventilation of these painted elements is very important. Eaves cornices, particularly, need care. If the roof is to be sealed for insulation this should be done between plate and roof covering, not at the cornice soffit, where the maximum ventilation possible (without admitting birds) should be given. No less important is the ventilation behind weatherboarding, behind the panelled reveals of door-cases, and behind such features as applied wood pediments and cartouches.

1.10 THE REPAIR OF MOULDED AND CARVED WORK

Beams and lintels

Moulded or carved beams, lintels and bressummers are often to be found in medieval work. The value and the interest of the carving may be such as to make its conservation essential at all costs even if the member is badly decayed. Provided the cause of decay is removed, it may be possible to carry the load a beam was intended to bear in some other way and thus leave the decorated timber merely to carry its own weight. The back of a moulded plank lintel, for instance, across a medieval window embrasure may be relieved of its load by a secret reinforced concrete lintel above, spanning not only the opening but the length of the lintel as well. In an extreme case where such a lintel is so weak as to be unable to support itself it can be hung from the relieving beam. Surface reinforcement or renewal of moulded members is clearly to be avoided; the preservation of the authentic original is of primary importance and remedial work such as scarfs and halvings must be so cut as to leave the moulded surface untouched. This may well lead to a situation where the scantling of the unmoulded portion is inadequate for the work it is intended to perform. A combination of methods may provide the solution. If it is a beam that is being repaired the added section may be reduced in depth if it is itself reinforced by the inclusion of a steel flitch. There can be no universal answer to such problems; they must be examined individually in the light of the principle that the original moulded work must be preserved while the member as a whole must continue to perform its designed function. In the last resort, and this inevitably means removal, the decorated surface can be cut off and used as a casing to a new built-up member of timber or steel. (See below.)

Cornices and built-up features

Cornices, both medieval and late, over doors and other such features were invariably built up of a number of members of small scantling, not only for ease of working but to allow for movement. It is important that in their repair this construction should be maintained and that the repaired work should have the flexibility of the original. As with the heavier structural timber it is the moulded surface which must be preserved and which often may be saved by carefully paring off the decayed back and remounting on new. The ventilation of the bracketing of such features is of great importance and full allowance must be made for this when they are refixed after repair. Similarly, every opportunity should be taken to isolate them from potential sources of future decay.

16

Preparation of a beam end to receive a steel flitch plate, using a traditional two-handled pit saw to cut the slot. (Photograph: DAMHB Archive)

1.11   THE USE OF SUBSTITUTE MATERIALS IN REPAIR

Steel in exposed timber beams

While, as an ideal, timber should always be repaired with timber, there are bound to be many cases where it is expedient to repair with steel which, in exposed timber, must be so placed as to be concealed when the work is completed. The method of most universal application is the use of a flitch plate (Figure 1.2). Here, the decayed or damaged portion of the timber is removed and replaced by new, butted to the old and joined by a vertical plate housed in a central slot and carried back at least three times its own depth on either side of the junction. The plate is secured by a series of staggered bolts whose heads are sunk below the surface. The bolt heads and underside of the plate, which should be made to allow for this,

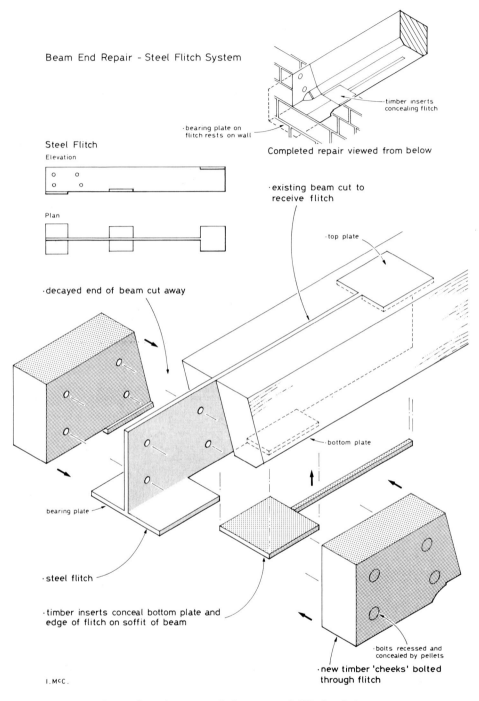

Beam End Repair - Steel Flitch System

·timber inserts
concealing flitch

·bearing plate on
flitch rests on wall

Completed repair viewed from below

Steel Flitch

Elevation

Plan

·existing beam cut to
receive flitch

·top plate

·decayed end of beam cut away

bottom plate

bearing plate

·steel flitch

·timber inserts conceal bottom plate and
edge of flitch on soffit of beam

·bolts recessed and
concealed by pellets

·new timber 'cheeks' bolted
through flitch

I. MᶜC.

Figure 1.2   Beam end repair using a stainless steel flitch plate
*The plate sits in the pit-sawn slot shown in preparation on page 17. The steel is
completely concealed, on completion of the repair, by oak cheeks and inserts.*

18

are concealed by wood slips. If the repair is to the bearing end of the beam, the plate will require stiffening with side angles bearing on the padstone or wallplate. If the end is visible, false sides may be built up in timber to conceal the plate, but if it is buried in a wall, only the plate should be built in. Side plates are unsightly and should be avoided where the repair is visible.

A fractured, as opposed to a decayed, beam may be repaired by restoring only the tensile element of its strength in steel, for the upper portion may still retain ample resistance to compression.

A small chase along the bottom of the beam, perhaps hidden by the removal and later replacement of a moulding, will be sufficient to house a tension wire or rod anchored in sound timber near the ends of the beam. This method has the advantage that it may be carried out with little or no disturbance to surrounding work. The support of a beam on a steel corbel need not be unsightly. At its end the full depth of the timber is not structurally essential. The corbel can, therefore, be recessed into a slot or housing provided that about two-thirds of the depth of the beam is maintained. In such ways as this it is always possible to avoid the surface use of steel on exposed timbers.

The use of resins and reinforcement

Epoxy resins have played a minor but important role in the *in situ* repair of timber over the past two decades. During this time techniques and materials have developed quite considerably, and there is no doubt that their use has made possible the retention of a significant amount of historic timber which, otherwise, would have been replaced. Unfortunately it is easy to use resins carelessly and it must also be said that disfiguring and incompetent work has been carried out which would have been much better accomplished using traditional timber-to-timber repairs. Resin repairs must not be used as a cheap alternative to craftsmanship; they must always be carried out by competent tradesmen capable of using traditional and new techniques in an intelligent and effective way.

The major structural uses of the epoxy resins are in the *in situ* repair of beam ends, the grouting and filling of timber sections excavated by fungal and insect attack and the *in situ* strengthening of floor beams which have become over-stressed due to cutting or overloading, or are required to take additional loading due to change of use.

Typical beam end repairs consist of providing temporary support, cutting away the decayed end, placing permanent wood formers to match the original wood and drilling through sound wood from the top of the beam at an angle into the bearing zone. Reinforcement rods are placed in the drillings, followed by a pour of an epoxy resin grout to fill the voids and permeate the semi-decayed area between the bearing and sound wood. The result is a bearing end of cured resin tied through resin-impregnated wood to sound wood. The wood forms can be adzed back flush with the old beam face, providing a completely secret repair with a new beam end which is no longer susceptible to wet or dry rot. Typical rods are formed of 65 per cent glass roving bonded together with an unsaturated polyester resin. The modulus of elasticity of these rods is similar to that of seasoned oak and they have a shear strength at least three times greater. Threaded stainless steel rods are also

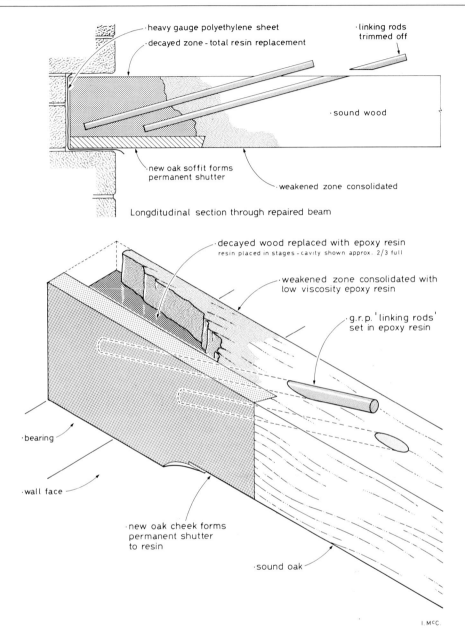

heavy gauge polyethylene sheet

·decayed zone - total resin replacement

·linking rods trimmed off

·sound wood

·new oak soffit forms permanent shutter

·weakened zone consolidated

Longditudinal section through repaired beam

·decayed wood replaced with epoxy resin
resin placed in stages - cavity shown approx. 2/3 full

·weakened zone consolidated with low viscosity epoxy resin

·g.r.p. 'linking rods' set in epoxy resin

·bearing

·wall face

·new oak cheek forms permanent shutter to resin

·sound oak

I. M^CC.

Figure 1.3   Beam end repair using epoxy resin and filler with linking rods of glass reinforced plastic

In this illustration a resin has been chosen to preserve the maximum amount of original timber. The resin is completely contained in oak formers which are a permanent part of the repair. Weakened wood is impregnated with resin and the rods link the resin end, through the reinforced zone, with sound wood. Repairs of this kind must be carried out by a skilled carpenter in just the same way as any other timber repair to a historic building.

used. Many hundreds of this kind of repair have been carried out in Britain, Europe and North America.

Floor beam strengthening techniques may involve cutting a slot along the beam length and inserting stainless steel rods bonded with epoxy resin or introducing a lattice of glass fibre rods set in resin along the beam length. These systems are to be preferred to unsightly plates mounted on the surface and may be much more efficient, providing a 30 per cent–50 per cent increase in the load carrying capacity, but they must only be used after a thorough structural analysis has been made and calculations have been made by a structural engineer. A company specializing in this kind of repair should be expected to employ its own engineer, who will supply the necessary analysis and calculations for approval.

Fillings are normally straightforward, but may involve low viscosity grouting and priming as well as low-density epoxy mortar packing. To avoid high exotherms the quantities of grout poured to cure at one time must be limited and on large jobs temperature checks should be made. Cured epoxy is not readily combustible.

Where dismantling is impracticable, it may be possible to carry out pinning and grouting *in situ* with rods set in resin. In this context it is worth noting that the Building Research Establishment carried out load tests on a scissor truss to compare the effects of load on pinned joints made rigid with epoxy repairs. A predictable reduction in deflection of rafter members was noted after the joints were made rigid, but bending stresses were reduced at the rafter ends. In the case under test the induced stresses were within acceptable limits, but the effects of making joints rigid and the consequent stress redistribution must be considered for every situation independently.

Reference is recommended to *Epoxies for Wood Repairs in Historic Buildings* by Morgan Phillips and Judith Selwyn and the APT's *WER Manual*, where particular specifications and design considerations in calculating loads and reinforcing systems may be found. (See References.)

The use of resins in surface filling

Wood fillers have a chequered history and have had a high failure rate. Haired lime, and gypsum, have been used in the past to make good large holes or shakes, and white lead and linseed oil on smaller areas. Both traditions were based on the need for the filler to have some flexibility to allow for seasonal movements of the wood. Modern proprietary fillers similarly seek to combine elasticity with durability. Among the most successful are two-pack 'glueing fillers' based on polyurethane resin binders and inert fillers. Applied with a filling knife to treated or untreated wood, these have permanent elasticity, good water resistance and good durability. In external situations filling must be especially carefully placed to assist the shedding of water.

Surface reinforcement

In the strengthening or repair of concealed timbers the use of side plates or other surface reinforcing is permissible. Care must be taken, however, to avoid placing plates against uneven surfaces, as the voids behind form exellent breeding grounds for beetles. If necessary, the surface must be trued to receive the plates. In all cases the steel must be properly cleaned and treated with rust inhibiting primer paint before being placed.

Fixings

The fixing at the point of repair and the jointing of the member as a whole must be differentiated, for the former must be rigid and the latter flexible. For instance, the bolting of plate to beam must be rigid and must be spread over a wide enough area to minimize the risk of splitting along the grain. Bulldog washers beneath heads and nuts are essential and bolt holes must be staggered to ensure the maximum distance between drillings. All sockets and voids should be sealed and grouted with glue. On the other hand, the bolting of the beam ends to padstones, wall plates or the connections between members should be through slotted holes to allow for seasonal or other movements. A twisting oak beam rigidly fixed at the ends can do considerable damage.

1.12 SELECTION OF TIMBER FOR USE IN REPAIRS

Secondhand timber

Where partial renewal or repair calls for the use of new timber in association with old, the ideal is to use reasonable secondhand material that has had an opportunity to age under similar circumstances to that which is being repaired. Minor defects such as old mortices and shakes need not be a problem, provided there is no doubt that the timber selected is sound and perfectly free from decay, and is approximately the right scantling, for even old timbers will twist on conversion.

New timber

If the right secondhand timber is not available or if there is separation between old and new (e.g. in the provision of one whole new truss among a series that are otherwise capable of repair), then new timber may, of course, be introduced; it must be selected with care. Firstly, the original type should not be varied, i.e. oak should replace oak and pine replace pine. Secondly, the quality should approximate to the original and, thirdly, the moisture content should be limited to 15 per cent, which is roughly the content to be expected in a standing building. As regards age, a useful rule of thumb is to use no timber that has not been in stick (i.e. air seasoned) in its required scantling, with allowance for working, for at least one year for every 25 mm (one inch) of thickness. If new timber is to be used in very close association with old, greater care must be taken to match its quality, grain and moisture content. This is particularly important where the new is to be bonded to the old with an adhesive; otherwise seasonal movements will produce twisting and distortion out of all proportion to that which might normally be expected. Carpenters in the past made generous allowances for flexibility and movement in timber structures, whether they were constructing a framed building or a run of panelling. In these days of higher internal temperatures it is advisable to be even more generous.

1.13 FINISHES—THE USE OF ADZE AND PLANE

The artificial ageing of exposed oak is not recommended. The early carpenters, who finished the surface of their timber with an adze, were highly skilled in its use and the surface they left may, at its best, be indistinguishable from a planed finish. While, therefore, the mechanically smooth surface produced by machine planing is undesirable, a scalloped 'antique' appearance is equally so. An adze should be used, but every effort should be made to produce an even surface with it. Combing with a saw blade or rasp file to exaggerate the grain may be used with discretion when matching in a small insertion, but should not be used in large areas. Machine finishing should be avoided at all times and clauses inserted in specifications to ensure that saw marks etc., on both soft and hardwoods, are removed by hand planing. This applies equally to machine-run mouldings, which can always be detected against those run by hand. It is best to machine oversize and finish on the carpenter's bench.

1.14 CLEANING OF UNPAINTED WOODWORK

The larger structural timbers need no treatment other than the use of a clear preservative after the removal of all dust and dry dirt, preferably by vacuum cleaner. Old timber should never be darkened artificially and in general it is unnecessary to stain new insertions, for the patch will, in due course, weather on its own. Stained or dirty woodwork may be successfully cleaned by simply scrubbing with neutral pH soap and warm, clean water. Old varnish may be removed by a solvent made up from American turpentine (not substitute) and acetone in proportions to be determined by experiment, followed again by washing down. Care must be taken to allow adequate ventilation and/or to wear a filtered mask. It is unwise to treat exposed woodwork, especially oak, with linseed oil which, being slow-drying, collects dust and dirt. The application of a clear proprietary preservative is advisable, to be followed, when the surface is quite dry, by a thick protective coat of clarified beeswax dissolved in pure turpentine finally rubbed down with a soft cloth. Alternatively, cleaning with solvent may be followed by a combination treatment of preservative (organosolvent type) in which the wax has been dissolved. Microcrystalline waxes may be substituted for beeswax at a suggested loading of 25 cc of wax per litre of preservative.

1.15 CONSERVATION OF WOOD SURFACES

Scope of treatable surfaces

There are very few instances where the surface of a timber can be judiciously ignored whilst deciding on its repair. This applies not only to carved and moulded woodwork, but to commonplace items such as floorboards or roof timbers. It has only recently been discovered that apparently haphazard saw marks across some timbers are important measuring points. Understanding these reference points could throw vital evidence on reconstructing the original form of altered buildings. Thus, automatically to cut back a rotten beam to where it is sound or even de-frassing without inspection to make a conventional repair with new timber could unwittingly destroy important information. Although this may seem unduly cautious it highlights the importance of studying the surface of commonplace timbers in historic buildings when deciding on repair.

The obvious situations where the surfaces are of overriding importance are where they are moulded, carved or painted, often a combination of the three. In these situations the conservation of the surface must be given equal prominence with any necessary structural repairs.

Instances where conservation of commonplace surfaces should influence considerations for any necessary structural repairs may be offered as follows:

- Single surviving elements of old structures, however humble.
- Plain elements which, through the passage of time, have a historical importance and may provide a link leading to our understanding of their similar or different use in the past.
- Conversely, individual pieces in a uniformly worn or aged object, where renewal could adversely upset an established balance.

The retention of problem surfaces must involve their conservation, and a conservator must be involved at an early stage. The type of conservation will depend on the function the conserved piece has to perform.

Different functions suggest different techniques: thus a projecting carving on a staircase newel cap receiving frequent handling may need different treatment from a carved cornice remaining untouched from one year to the next.

Materials for consolidation

There seems little evidence of past efforts actively to conserve surfaces through consolidation and strengthening. Shellac has been used for the last 200 years or so, both as a varnish and, on hidden surfaces, to seal against damp; waxes have traditionally been used for surface protection, and recently, and very damagingly, a liberal application of linseed oil has been widely used to enhance old dusty surfaces. In the 1920s, the SPAB was recommending the use of magnesium silicofluoride. None of these materials has many of the properties one would currently consider desirable: long term stability in wet or dry conditions, clarity, non-yellowing, non-staining, flexibility, high viscosity, and long term stability. Reversibility is highly desirable, but if a particular consolidant is chosen as clearly the best for the job, a less satisfactory material should not be chosen just because it is reversible. In a situation where two consolidants are equally satisfactory, the more easily reversible should be chosen; in some circumstances, reversibility might in fact be of primary importance.

There are a large number of possible consolidants, but at the present time the two groups which seem to be most useful are the acrylic and epoxy resins. Both fulfil many of the desirable characteristics; both have flaws, but each has certain characteristics which make it more suitable for specific purposes.

A timber sculpture covered with gesso and polychrome and substantially weakened by wood-boring larvae is strengthened by painstaking injection of a low viscosity resin under the decorated surface. There are no short cuts to this kind of consolidation, which must be carried out by a trained conservator. (Photograph: Nicola Ashurst)

Of the two, the epoxy resins normally impart greater strength and more uniform consolidation, but are somewhat more difficult to introduce to great depths, and darken the surface. At times, too, their very strength can be a disadvantage. The acrylic resins tend to be highly viscous, do not darken surfaces and are reversible, but can produce uneven consolidation with the greater deposit of acrylic near the surface, and are not as strong as the epoxy resins. This last characteristic is considered an advantage in some situations, as the consolidated wood should have a degree of flexibility, and therefore be more in sympathy with adjoining untreated wood. These are very general comments, because properties such as viscosity and flexibility are infinitely variable, and even darkened surfaces can be treated so that there is no colour change when compared with untreated surfaces. New materials and techniques are being developed all the time, and the exceptions and special considerations with each job stress the importance of employing a conservator to help select the correct method and material for each part of the work. It is important to target individual treatments to individual problems, always selecting the least interventive and most easily reversible to do the job.

Considerations before consolidation

Any inspection leading to consolidation needs to provide answers to the following questions:

- Will the area to be consolidated be buried in damp conditions with or without ventilation?
- Is the area in a load-bearing position?
- Is the area in a frequently handled situation?
- Is the area in a high wear area?
- Will a colour differential after treatment be acceptable/unacceptable?
- Is the area to be treated isolated or occurring elsewhere?
- Is the cause of the decay still present or has it been removed?
- Is the object of great historical or artistic significance?
- Is the cost of treatment important?
- What access is available for treatment?
- If the decay is in a veneered or laminated structure, will the glue affect penetration of the consolidant?
- Is the surface to be cleaned as part of the conservation work?
- Is new wood to be fitted to the consolidated area?

Use of this checklist will help to anticipate problems and to prepare sensible specifications with realistic costs.

Treatment

On the assumption that the subject will be properly ventilated and reasonably dry, treatment of individual areas of wood can be undertaken without the same concerns that exist when treating stone. Consolidation and treatment of each area of decay is only of that particular area, and treatment of surrounding timber is not being sought. Sound timber is unlikely to benefit from the introduction of a consolidant as distinct from a preservative, and is unreceptive to both in site conditions.

However, the importance of thoroughly consolidating each area of decay cannot be too strongly stressed. It is vital that the whole decayed area is consolidated, and that the intermediate zone between decayed and sound timber is thoroughly impregnated with consolidant. Lack of consolidant in this area will result in a mass of consolidated material separated from the parent body by a layer of weak degraded wood. This is a very dangerous situation as, although the object may look and feel solid, minor knocking will detach the whole section.

Techniques

The consolidant should be fed in by brush or hypodermic syringe until no more will be taken up. If necessary, drill into flight holes of wood-boring insects to the full depth of the decayed area, using bit sizes as close to the holes' size as possible. Always wear protective clothing, paying particular attention to safety glasses, nose and mouth protection and surgical gloves, and ensure that smoking is banned in the work area.

The following points will be useful to remember:

- In cold conditions, resins will take two or three times as long to cure as in summer; rate of cure is temperature dependent.
- Resins when mixed and kept in deep containers will heat rapidly and may explode.
- Any excess resin must be cleaned off with solvents as rapidly as possible.
- Surface darkening can be reduced by cleaning the surface later with methylene chloride.
- A constant check must be kept for runs in all possible places.
- Areas not to be treated must be masked off or protected with a dammar wax coating.
- Consider the solvents which are to be used for later cleaning. Acetone used for cleaning will rapidly strip any acrylic resin, for example Paraloid B72.
- Consider whether to shape a surface for fixing new wood to consolidated area before consolidation or after.
- Consider which glues are to be used for fixing and grouting new wood.

Consolidation to one side of a pierced carved panel or board might easily cause it to warp, both after application of consolidant, and later when unequal intake of moisture between front and back may cause differentials in movement in the wood. Both sides may need treatment to ensure stability.

Epoxy resins are quite heavy, so that new stresses may be caused by consolidated areas being supported on inadequate stems of old wood. Allowance may need to be made for strengthening support zones.

In general it is recommended that lacunae or flight holes are made good with consolidant/wood filler. The extent of a filling can always be detected in the future, and at the present time no cutting away to splice in new is necessary. The fillings should be in visual and physical harmony with the treated timber in which they are made, keeping the amount of resin as low as possible consistent with the workability of the filler. The filler should be kept at or just below the original surface levels and toned with pigment to the colour of the surrounding wood. Large batches of epoxy resin and filler must not be made up at one time because of the exothermic problem.

Painted surfaces

Painted surfaces present particular problems for the conservator which are additional to those already described.

Apart from mahogany, nearly all internal decorative timber, such as panelling, fireplace surrounds and doors, was originally painted. Staircase handrails, some balustrades and marquetry panelling are obvious exceptions, but the general rule was that they should be decorated. Panelling, for instance, in the seventeenth and eighteenth centuries, whether of oak or pine, was frequently painted and grained to represent various rare woods or marble. Careful search should be made for this before redecorating and, if not recoverable, it should be reproduced. The fashionable stripped pine panelled room is unlikely to be historically accurate and is to be discouraged. In medieval work paintings may be expected on beams, in the hollows of mouldings, on wood ceiling panels and in the panels of screens. In later periods, particularly among bolection moulded work, they may appear on over-door and overmantel panels, perhaps overpainted to match a later scheme of decoration.

It is clearly undesirable to consolidate dirt onto a fragile original paint surface, and a consolidant injected behind paint surfaces may seep between original paint layers and later layers making cleaning off of the latter to reveal the original almost impossible. Some paint and pigments, mostly with glue medium, can be irretrievably damaged by consolidation with the wrong material.

In some situations the paint surface will have to be cleaned and consolidated with a suitable consolidant before the wood behind can be treated.

When wood-boring insects tunnel right behind the paint layer it becomes very vulnerable to damage and loss, even though the surface may look quite solid. The problems of injecting consolidants into these voids is currently so time-consuming and so potentially damaging to the surface that this work should only be attempted where actual survival of the object is at risk.

In general, great caution must always be taken with paint, and a polychrome conservator should always be involved. This also applies to old varnished surfaces. These have almost always been refreshed with further coats of probably different varnish. The greatest care should be exercised in cleaning if the retrieval of the original surface is to be achieved.

Consolidation of external surfaces

So far, little work has been done in this field. Deep shakes and loosening of knots occur more readily with the greater variations in temperatures and humidity, and surface checking also occurs, leading to deep pitting of the surface and decay around water traps. Whenever possible these should be filled to prevent retention of moisture and, indeed, to encourage run-off of surface water wherever possible.

Both acrylic and epoxide resins may be used to grout and fill, although the acrylics have generally been found to be more satisfactory. When piecing in new oak it must be remembered that it will open up along the lines of medullary rays. The grain of the oak should match the direction of the original grain and shed moisture, not collect it. Letting in small pieces of timber should be avoided as far as possible, since the differentials in movement result in the new pieces coming loose quickly and either dropping off or actually trapping water behind and increasing the rate of decay.

Surface consolidation and surface repairs

The kind of structural repairs described above may well be to decorated wood sections. In this case surface conservation techniques must be applied first to retain as much of the original material as possible by consolidating the surface and cutting out decayed timber behind it later. Consolidation of surfaces should not be carried out when untreated timber behind is left in contact with damp masonry, because decay may then proceed unnoticed through the core behind the consolidated surface. Beam failure in this case may be a considerable archaeological as well as structural loss.

Conclusion

As much of the original surface must be kept as possible, even when it is weakened and frass is present. New techniques are available using wood consolidants. As is always the case when introducing synthetic chemicals into natural materials, the greatest care must be taken. No short cuts in application are tolerable; they would almost certainly lead to worse problems in the future. Techniques are laborious and slow, and therefore usually quite expensive. Consolidation should not be seen as a cheap alternative to conventional repairs. The saving of original detail is, however, of incomparable value.

It must be strongly emphasized that the cleaning and treatment of important surfaces is the work of professional conservators.

REFERENCES

1 Association for Preservation Technology *The Wood Epoxy Reinforcement System Manual* (WER), APT, Canada, 1979.
2 British Standards Institution:
   BS144: 1973, *Coal Tar Creosote for the Preservation of Timber.*
   BS1282: 1975, *Guide to the Choice, Use and Application of Wood Preservatives.*
   BS3452: 1962, *Copper/Chrome Water-borne Wood Preservatives and their Application.*
   BS3453: 1962, *Fluoride/Arsenate/Chromate/Dinitrophenol Water-borne Wood Preservatives and their Application.*
   BS4072: 1974, *Wood Preservation by means of Water-borne Copper/Chrome/Arsenic Compositions.*
   BS5056: 1974, *Copper Napthenate Wood Preservatives.*
   BS5268: Pt 5 1977, *Preservative Treatments for Constructional Timber.*
3 Berry, R W , *Cyperethrin: A New Insecticide for Wood Preservation*, Building Research Establishment Information Paper.

4 Building Research Establishment:
   Digest 175, *Choice of Glues for Wood*, March 1975.
   Digest 201, *Wood Preservatives: Application Methods*, May 1977.
   Technical Notes:
   No 39, *The House Longhorn Beetle*, March 1969, reprinted September 1980.
   No 7, *Insecticidal Smokes for Control of Wood-boring Insects*, February 1966, reprinted April 1980.
   No 47 *The Common Furniture Beetle*, November 1970, reprinted June 1980.
   No 45, *The Death-watch Beetle*, October 1970, reprinted May 1979.
   No 44, *Decay in Buildings, recognition, prevention and cure*, December 1969, revised January 1977.
   Building Research Establishment, Watford, England.
5 Phillips, Morgan and Selwyn, Judith, *Epoxies for Wood Repairs in Historic Buildings*, Heritage Conservation and Recreation Service, Technical Preservation Services Division, US Department of the Interior, US Government Printing Office, Washington DC, 1978.
6 Richardson, Barry A, *Wood Preservation*, The Construction Press, Lancaster, 1978.
7 Richardson, Barry A, *Remedial Treatment of Buildings*, The Construction Press, Lancaster, 1980.

2 THE REPAIR AND MAINTENANCE OF HISTORIC GLASS

Jill Kerr

Jill Kerr, MA, was Secretary of the Corpus Vitrearum Medii Aevi Great Britain for ten years. She is now a liveryman of the Worshipful Company of Glaziers, a member of the Glaziers Trust and represents the HBMCE on the Council for the Care of Churches Advisory Committee on stained glass. Miss Kerr is an inspector of ancient monuments and historic buildings for HBMCE where she deals with grant aid to churches and coordinates advice on the conservation and repair of historic glass.*

2.1 SCOPE OF THE CHAPTER

Glass is a complex man-made, non-crystalline material. The basic ingredient, natural silica in the form of fine sand or flint, was fused with alkaline fluxes such as soda and potash and other ingredients including lime and cullet (broken glass). After the molten glass was formed to the required shape it was annealed, cooled slowly and evenly to strengthen the glass by removing the stresses that had built up during manufacture. The glass could then be cut.

This chapter considers the conservation, repair, maintenance and protection of window glass in historical buildings.

The technology of glassmaking existed in Europe 4,000—5,000 years ago and was introduced to Britain by the Romans. The history of glass manufacture in Britain is extensive and as it is not possible to summarize it simply, it is not included in this chapter. Further information can be obtained from the references listed at the end of the chapter.

*The author gratefully acknowledges the professional and technical advice contributed by Ian Curry, Harry Fairhurst, Alfred Fisher and Alan Younger.

Glass is a very vulnerable and brittle medium which decays due to factors which are inherent in it, external to it or a combination of these. A useful list for further reading into the complex mechanisms of the corrosion of glass can be found in reference 3.

Professional advice and craft skills

Those who work on or make decisions regarding historical window glass must understand the complicated nature of the material itself; the way it is fitted into the window/building context; the ways in which it interacts with its environment and the combined effect of these. Architectural advice, desirable for an overall view of the performance of glass, should be sought from specialist professionals who have worked within the large and small scale vagaries of historical window glass. The conservation and glazing skills necessary for undertaking work on the glass, leads, ferramenta and window surrounds should also have been proven. The Council for the Care of Churches has a list of specialist glaziers but, as glazing should not be considered in isolation to its building context, it is usually advisable to seek such advice through an architect who will have knowledge of the associated masonry, building and non-glazing problems.

2.2   GLOSSARY OF TERMS

'Antique' glass: Common term for hand-made glass blown by the 'muff' method (see below) and thus containing bubbles, ripples and irregularities.

Backpainting: Painting on the exterior surface of the glass. Yellow stain is usually applied to the exterior.

Came: The strip of lead, H-shaped in cross-section, which is wrapped round the edge of the cut glass pieces to make up the design panels. The cames are joined by soldering, and the gaps between the flanges and the glass surface are filled with putty or mastic to prevent water penetration (see 'cement').

Cartoon: Full-size design for a window complete with lead lines and detail to be painted.

Cathedral glass: Commercial name for machine rolled glass.

Cement: Technical glazing term which refers to the putty or mastic filler between the glass and the cames which links them and makes the window waterproof.

Crizzling: The roughening, crumpling or scaling of the surface of glass which impairs its clarity. A form of decay and decomposition associated with fire damage.

Crown or spun glass: A bubble of molten glass which has been removed from the pot in the furnace, is opened at one end and spun on a pontil rod so that the centrifugal force flares the glass out to form a flat disc with a thicker central pontil knob or 'bulls eye'.

Cullet: Broken glass, recycled and used in the production of new glass.

CVMA: Corpus Vitrearum Medii Aevi

Dalles de verre: Thick slabs of glass produced by casting in moulds. As the slabs are too thick for retention in lead, they are set in concrete or resin and chipped or faceted to the required shape (20th century).

Ferramenta: The iron framework or fittings which provide a fixing for panels within a wide window space. The panels are held in the frame by triangular metal pegs inserted into lugs on the main frame. The design of the ferramenta is an integral feature of the panel arrangement.

Flash: A thin coat of coloured glass fired ('flashed') onto the surface of white or pot metal glass during manufacture. The flashing can be abraded or removed by acid to reveal the colour of the base glass, achieving two tones on the same piece of glass, a technique frequently used in heraldry. Red or ruby glass is commonly flashed.

Glazing bar: A horizontal or vertical T-shaped support for panels of glass in large openings. The glass rests on the T-shaped profile. Sometimes called a T-bar.

Grisaille: Geometric or leaf patterns of regular design leaded into or painted on white glass. (From the French *grisailler*, to paint grey.)

Grozing: The medieval method of shaping glass by means of a metal tool with a hooked end which made a characteristic bitten edge. (Post-medieval glass is cut by a diamond or sharp metal tool, which creates a straight edge.)

Muff or *cylinder glass*: An elongated balloon of molten glass is blown. The round ends are cut off to form a cylinder shaped like a muff which is then cut along its full length and relaxed in an annealing chamber until it becomes a flat rectangle.

Pot metal: Glass coloured throughout with one or more metallic oxides when molten in the pot.

Quarry: A square or diamond-shaped pane of glass, usually white. Plain, unpainted quarry glazing is generally made up of diamond shapes which serve to carry the combined weight of glass and lead vertically without the distortion frequently encountered in the horizontal stress pattern of square panes. (From the French *carré*, square.)

Tie bars: Panels within ferramenta or T-bars are held in a stable position by being tied at intervals to cross bars. The tie of lead or copper is soldered on to strong points on the cames and attached firmly to the bar. These are sometimes called 'saddle bars' and can be round or square sections. There is some advantage in using a square section as there is less likelihood of the copper tie being torn loose from the solder when tightened.

Yellow stain: A stain ranging from pale lemon to orange, produced by applying a solution of silver compound to the surface of the glass which, when fired, turns yellow. Sometimes called 'silver stain'.

2.3 CONSERVATION PRINCIPLES

There are three basic principles for the conservation of all types of historical glass. These should be born in mind whenever work is proposed or undertaken. 'Conservation' of historical glass means that every attempt is made to keep as much of the original glass, lead, ferramenta and the surrounding masonry as possible.

The conservation of historical glass should always encompass the principles of minimum intervention, full recording (before, during and after the work) and the use of reversible techniques.

This section of glass would greatly benefit from a gentle clean to remove disfiguring superficial dirt. The desirability and feasibility of removing the mending lead which cuts through the original design of the three heads should be investigated, as well as appropriate crack mending techniques. These decisions and works should only be undertaken by trained and experienced conservators and craftsmen. (Photograph: Jill Kerr)

Minimum intervention

Current practices of glass conservation are based on the principle of doing as little as possible to the glass. Some glass can be cleaned *in situ*, some without stripping the leads. It is important that the expectation of dramatic transformations of glass is not fostered. Most conservators will advocate the removal of harmful and damaging dirt and corrosion products, the increase of legibility by the removal of lead lines destroying designs, and the replacement of missing features using backing plates. The rearrangement of ancient glass and the removal of subsequent additions and repairs is not encouraged, as the arrangement of the glass prior to conservation is part of its history, and intrinsically worthy of preservation. Rearrangement such as right-siding reversed glass is usually only carried out if the integrity of the overall arrangement is not seriously jeopardized. Speculative and gratuitous rearrangements have destroyed a significant amount of historic evidence and must be avoided. The reconstruction of lost iconography and the restoration of designs such as heraldry should never be attempted without the backing of documentary evidence, and the collaboration of a specialist iconographer/glass historian. The 'leave as found' principle is the primary concern of current conservation practice. Any departure from this principle must be fully argued, justified and recorded.

Full recording

A full record should be made not only of the state of the glass before work commences, but also of every detail of every process used during conservation. The pre-conservation record must include diagnosis and analysis of the precise nature of the decay and the extent of the damage on both surfaces of the glass. Photographic records in colour and black and white should be made of the interior and exterior of both surfaces of the glass, with macrolens details of specific problems to supplement the written account. A pre-dismantling lead rubbing should be made. This can be photographed or photocopied for easy reference to record every stage of the conservation process as it takes place on the bench. The conventions for recording the above procedures are reproduced on pp.65–69 from the recording system of the Council of Care of Churches and the International Corpus Vitrearum. It is essential not only that each practising conservator should know precisely what he is doing and why he is doing it, but also that future conservators should know what has been done. It is only by learning from past failures that techniques can be refined, developed or discarded. One of the major problems glass conservators have inherited from the past is inadequate records of previous methods, and valuable time and resources have been diverted to researching these. When the conservation work is complete, the panels should be photographed again in the same detail. The CVMA/CCC method of recording is an accepted part of all grant-aided glass conservation and the cost of it is included in the conservator's estimate.

The importance of keeping dated records of glass cannot be emphasized strongly enough. They often provide invaluable evidence of a previous state of a window thereby enabling appropriate remedial work to be undertaken on it. Photographs provide a record of the condition of the glass and its surrounds which in turn provides important comparative evidence of rates of decay. Glass is almost always at risk from some threat or other and requires the precaution of proper recording, even if no work is to be done on it.

Copies of records can be sent to the CVMA Archive at the National Monuments Record where advice can also be sought on all aspects of recording glass *in situ*, in collections or in auxiliary material such as cartoons or drawings presented by the artist or restorer to his client (see p.64 'Sources of Further Information'). It is important that all stained and painted glass is recorded wherever it is and in whatever condition.

Reversibility of techniques

It is a paramount principle that no techniques are used in glass conservation that are not reversible. This eliminates many of the damaging methods used in the past, such as the application of chemical coatings, the refiring and repainting of historic glass and the destruction of historically important arrangements. No reversible techniques have so far been developed for the fixing of loose paint.

Exchange of professional experiences

The three principles of full recording, minimum intervention and reversibility detailed above are the basic rules to follow in the cleaning and repair of all types of glass. It is always a mistake for conservators to operate in isolation from the international community of specialists who are continually confronting the same technical and professional difficulties. Sharing experience and knowledge, success and failure is an essential part of professional conservation. It is rare to encounter a problem in conserving glass of any type or period which has not been experienced by another specialist from whom much could be learned. This is particularly important when dealing with foreign glass, where communication with the experts in the country of origin is essential in keeping up-to-date with the continual evolution of techniques and methods. It is also useful to build up a network of contacts among colleagues with whom it is both encouraging and informative to share experiences. Always, if possible, attempt to see a new technique or material in action before deciding whether to use it.

2.4 THE INSPECTION AND CONSERVATION OF LEAD CAMES

Lead cames hold the glass pieces together in such a way that the window is flexible and able to respond to expansion and contraction in reaction to daily and seasonal temperature changes, as well as temperature differential between the inside and the outside of the building. Most important, leads are also historical documents and should be considered an integral part of the design.

In the Middle Ages, the glass designer minimized the inconvenience of having to change the glass every time he needed to change colour by carefully integrating the cames into the design. It is often forgotten that we are frequently looking at medieval glass through a disfiguring network of supervening mending leads

Medieval window lead profiles (cast)

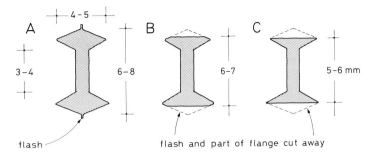

Cames cast in hinged two-piece moulds about 500 mm long.

Type A has thick diamond shaped flanges and a prominent casting flash along the outside edges.

Types B and C have been made from cast came (as Type 1) by scraping off the casting flash. Types 2 and 3 differ only in the amount of lead removed from the flange. Being hand-made there is considerable variation in Types A, B and C, even within the same piece.

Post-medieval window lead profiles (milled)

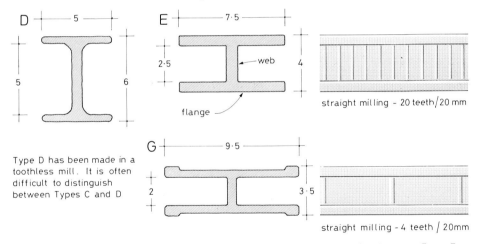

Type D has been made in a toothless mill. It is often difficult to distinguish between Types C and D

There appears to be no clearly defined boundary between Types E and G; the flanges of Type E become wider and thinner and the tooth count decreases from 20 teeth in 20 mm to 4 or 5. The web of type G is sometimes inscribed with the maker's name or initials and date.

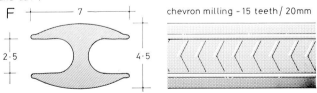

Figure 2.1   A chronology of the profiles and methods of manufacture of lead cames in medieval and post-medieval windows (Dr Barry Knight)

inserted by subsequent repairers, restorers and jobbing plumbers. This point was often misunderstood by the glass designers during the Gothic Revival who frequently produced glass without integrated lead lines, following their own design rather than the logic of the deployment of colour and line the medieval glazier would have used. It is possible for an experienced conservator of glass actually to restore the legibility of incoherent medieval glass by reconstructing the original cut lines. It is also possible in some cases for a glass historian to recover lost iconography by studying the surviving lead pattern.

A rubbing of the lead pattern of each panel should be made before the glass is dismantled. This serves as the basis for the record of the restoration itself and provides the working cartoon for the reconstruction of the panel and the laying out for reglazing. Again, it is important for the leads to be examined in detail so that any historical evidence for the dates, and the extent of subsequent repairs and releading, can be gleaned from the leads themselves. In some cases leads were inscribed, signed and dated, and much evidence for the development of lead-came technology can be derived from a detailed survey. As all lead cames eventually relax and deteriorate due to continual movement, all glass requires releading at intervals in its life. Because of this, the survival of medieval cames is very rare, which makes their recognition, retention and recording even more important. Present principles of conservation favour the retention of medieval lead wherever possible, the glasses they hold being cleaned *in situ* and any strengthening necessary being carried out with sympathy and restraint. At the present rate of releading of sixteenth-, seventeenth- and eighteenth-century glass, cames of this period will soon become equally rare.

2.5 FERRAMENTA AND FIXING METHODS

A miscellany of methods of fixing glass into glazing spaces has been, and continues to be, used for all types of window glazing. Some of these are of particular historic importance as they can provide essential evidence for the reconstruction of lost glazing programmes and should not be removed or destroyed.

This is particularly so in the case of medieval ferramenta where the survival of the armatures, the iron infrastructure constructed in geometrically symmetrical patterns to whose frames the panels of original glass were pegged in place, can survive intact, glazed with nineteenth-century reticulated white quarries. While this is a ghost of the medieval glazing programme, it can be indicative of the number of scenes selected from, for example, the life of the saint whose lost altar below is recorded in antiquarian sources. Such survivals generally occur in cathedrals such as Canterbury and Ely, but equally interesting evidence for early examples of secular glazing and window fitments can be clearly observed in the frames of many buildings. Hinges for shutters, rebates for the insertion of wooden frames containing horn or oiled cloth, chamfered glazing grooves (designed to take the precious glass with which the rich magnate or bishop travelled, to embellish his apartments or proclaim his status in his audience chamber), can be just as interesting to glass historians as the humble, original metal fittings that survive in the window frames of a yeoman's cottage. Nothing should be destroyed or removed without knowing what it is, its function and date, its rarity value, its potential for re-use and the extension of its life, or before it is recorded by measured drawings or photographs. This is particularly important for obsolete survivals such as the ancient oak frame into which the glass was nailed and then rebated into the masonry, and for medieval iron bars which may assist with determining the original function of a room.

Non-ferrous tips can always be added to ancient tie bars, or metal grilles, so that rust disruption of the stonework in which they are fixed is prevented. The techniques of preserving corroded iron are well established (see Volume 4, Chapter 2). A blacksmith working with a specialist conservation glazier can retain and restore or copy decorative features such as casement hinges and handles.

An important detail often overlooked is the composition of the mortar bedding with which the glass is glazed into the groove of the masonry frame or into which the hinge is fixed. If the mortar is too hard, badly placed so that it spills over onto the glass beyond the lead edge flange, the wrong colour for the adjacent stonework, or inexpertly applied, it can not only disfigure the building seriously but also be a source of damage to both glass and stonework. Dense cementitious mortar fixings also create difficulties for future deglazing which will at some time be necessary. The removal of such mortars may require the use of potentially damaging techniques such as mechanical grinders. If it is undertaken in an inexperienced manner both the stone and the glass may be damaged. This is particularly so when there is no sacrificial framing or white glass fillet at the edges of the glass. The architect should specify the constituents, proportions and technique of the mortar and identify any physical problems relating to the fabric and the architectural context in which the glass is set. If necessary a mason or a blacksmith should be brought in, under architectural supervision.

2.6 THE DIAGNOSIS AND TREATMENT OF GLAZING DEFECTS

Glass is a vulnerable and brittle medium which should be inspected frequently to ensure that any changes, however undramatic, are observed and recorded. Even small changes should be reported to a specialist architect or a conservator/glazier, with sufficient clarity for them to decide whether or not a detailed inspection is necessary.

Historical glass can be adversely affected by many factors external to it. It can, at times, become linked into the transfer of the loads of a structure to the ground. As the weakest element it will be the first to exhibit signs of distress, indicating that rectification of a structural problem is also necessary.

The following defects are the most commonly experienced in glazing panels. The causes of these and appropriate remedial actions are discussed. Details of the remedial procedures are included in following sections.

The deformation of the lower section of this window has resulted principally from the use of horizontal and vertical divisions. In this orientation the leads have been unable to sustain the vertical loads imposed on them. This type of failure rarely occurs with the diagonal leads of diamond quarries which transfer similar loads without distortion, hence their wider use. (Photograph: Jill Kerr)

Undulating or buckling panels

The principal causes of undulating or buckling panels are failure of the leading, inadequate tying or support, oversized panels, too much movement, the panel being too rigidly glazed or fixed without adequate allowance for expansion and contraction, and slippage of the support system. The presence of too many horizontal leads in a window can result in structural failure as leads in this orientation are easily overloaded (see above).

The cause or causes of buckling should be correctly identified and remedial works proposed under architectural guidance. Releading will only be necessary if the leads have failed. Deformed panels can often be gently flattened and re-installed with additional support. Replacement tie-bars should be non-ferrous or have non-ferrous tips.

Paint loss, lost enamel

The loss of paint and/or enamel can be caused by underfiring, the use of borax as a reducer of the melting point at which the paint fused onto the surface, inadequate flux preventing the enamel adhering to the glass surface properly, chemical or abrasive cleaning either on or in the vicinity of the glass or storage of the glass in an unsuitable environment.

Details of the surfaces should first be recorded in colour and black and white photographs using transmitted and surface light. The causes of the problem should be confirmed in a specialist glazier's report and a solution proposed which does not involve repainting or refiring of the original glass.

Replacing lost paint on backplate is appropriate only where a few important elements of the design are affected. It can be used selectively for faces, hands and other major features but is too expensive and heavy to use for all painted areas. It is a reversible method. Isothermal glazing is advised for more extensive areas of paint loss or for total failure (see p.59). This solution is a 'holding operation' to ensure no further loss through condensation, temperature extremes, weathering action etc. Sometimes the only viable solution to the problem is to record and leave.

There is no current reversible technique for fixing loose paint.

Organic growth on glass

Glass surfaces which are covered in weathering products, or are scuffed, or whose paint has a rough surface are able to retain sufficient nutrients to encourage the growth of disfiguring lichens, algae or fungi.

Such surfaces should be cleaned extremely carefully without damaging any loose paint on either surface. Only neutral-pH soap available from conservator suppliers should be used. Chemical cleaners, especially hydrofluoric acid, must never be used. Any biocide which is used must be 'safe' for all the materials with which it will come in contact (see Volume 1, Chapter 2 'Control of Organic Growth' for further details). Cleaning historic glass is a specialist operation.

Cracking, crazing or fissuring and fire damage

When the surface tension of glass is broken the crystalline matrix of the body disintegrates and various types of crazing, cracking or fissuring occur. This can be due to exposure to high temperature in a fire, or damage from misdirected directional heating devices.

The depth and extent of the crazing should be determined. If it is found to be superficial, the surface should be investigated for the application of a chemical or plastic coating or varnish at a previous time which may have failed or become damaged by heat or light. Characteristic crizzling combined with sugary opacity can be the result of too much alkali in the making of the glass. If the surface is sticky to the touch and exudes moisture, the glass was made with excessive potash as the alkali agent, and the surface is now composed of potassium carbonate which is hygroscopic and attracts and retains moisture.

The need for a specialist glazier's report to determine the exact cause and form of deterioration should be discussed with an architect. The solution may simply be to divert directional heating, or relocate a bonfire area outside the building. If the cause is failure of a chemical or plastic coating, this must be identified before any attempt is made to remove it. If fire damage is the cause, expert advice should be sought immediately. Fire crazing usually results in complete disintegration of the glass as soon as any dismantling of the cames is undertaken. No attempt should be made to touch or clean the surface or the glass however smoke-blackened it may be. A glass conservator should examine it and determine the extent of the damage so as to devise the best way to protect and remove it for conservation, if this is possible. Badly damaged glass and failed potash 'sugar' glass should be recorded *in situ* before replacement. It is not advisable to use either sticky tape or any proprietary brand of glue to consolidate crazed glass, as these can create more problems than they solve and are not always reversible or removable without extensive damage.

Shattered glass — shotgun or missile damage

A window which has inadequate exterior protection may be shattered by accidental damage, deliberate vandalism, stones flying up from lawn mowers, branches of trees or powerful winds.

In the event of such damage, every fragment from the broken window should be collected and loose or unsupported pieces that might subsequently fall should be removed. Whenever possible, the original position of the pieces should be noted. Sticky tape or proprietary glues should not be applied. All collected broken pieces should be retained for the conservator. The damage should be photographed and a photographic enlargement of any damaged area should be provided to assist the conservator. The insurance company, police and the architect may all need to be informed. Branches or bushes that threaten to beat against the glass should be cut back. Stones should be removed before mowing in the vicinity of vulnerable glazing. Missiles such as piles of stones or bricks should not be left to tempt vandals. Architectural advice should be sought on appropriate protective glazing or mesh. If vandalism is persistent, the police should be consulted on deterrent lighting, pressure pad alarm systems, restricting access, surveillance etc.

Water penetration

Water will penetrate glazing if the cementing medium between the cames and glass has failed, there are gaps in the glazing groove at the edge of the glass resulting from a failure of the mortar or the edge flanges of the lead panels, the overlap of the lead flanges between panels is inadequate or if the glass is cracked. Condensation on the inside of glass can be another source of excessive water (see below). Observation of a window during a period of rain can often reveal how water is penetrating. If no obvious cause of penetration is discernible, architectural advice should be sought. The window will require reglazing if it has to be re-leaded, re-weatherproofed or re-cemented. Depending on the extent and seriousness of the damage, most of the other faults can be rectified *in situ* by an experienced glazier, but it is he or she who should advise on the correct degree and method of repair. Rags or gobbets of putty should not be used to block holes, plastic or sticky tape should not be stuck onto broken glass and proprietary brands of filler or sealant should not be used to stop leaks.

The central design in this glazing, viewed from the exterior, is comprised of a backing plate originally installed to protect the inner medieval glass. The glue used to repair cracks in the medieval glass is prominent because it was not cleaned off the joints properly and an inappropriate glue was used. It was also unfortunate that the seal between the protective backing plate and the cames was incomplete as water has penetrated between the plate and the medieval glass resulting in the accumulation of disfiguring dirt. The correct procedure for back plating is described in section 2.8 'Repair methods'. (Photograph: Jill Kerr)

Condensation

Condensation is a deposit of moisture from the air onto the surface of the glass which occurs when the temperature of the glass is below the dew point temperature of the air. The diagnosis of the cause of condensation is not always straightforward (see Volume 2, Chapter 1 'The Control of Damp in Buildings'). Significant factors are often the water vapour in people's breath, moisture-emitting heat sources, the pattern of heating cycles, insufficient ventilation, the outside air temperature (seasonal changes) and the orientation and thermal mass of the building.

Appropriate remedial action of the rectification of a condensation problem may be as simple as improving ventilation or changing an unsuitable heating system, but it should always be discussed with an architect.

2.7   THE CLEANING OF GLASS

Every method devised for cleaning glass, including water and gentle brushing, can be damaging if the condition of the glass or paint is fragile or unstable. Before deciding upon a particular cleaning method, a detailed examination of both surfaces of the glass is essential. It is important that the reasons why a particular cleaning method is selected are recorded. Secrecy and the use of secret recipes can damage both the glass and a conservator's reputation. It is important to remember that nearly all the glass-cleaning techniques currently in use are at an experimental stage and many have actually accelerated the rate of decay of the glass on which they were used and created more problems than they solve in the short term. Controversy and criticism from glass conservators and technologists surround virtually every cleaning process.

Cleaning methods to be avoided

The following statements give guidance on how to avoid bad cleaning practice:

- Do not use any form of cleaning if loose paintwork or enamel have been detected on either the interior or the exterior of the glass. In that case the best course of action is immediate recording, assessment of the cause of the loss and detailed discussion with the client on a preferred strategy. Paraloids and wax have both been used to refix loose paint on glass which is to be replaced into 'museum conditions'. These fixing methods are very expensive and time consuming and should be avoided if glass is to be reglazed directly into a window.
- Detergents, bleaches, caustic sodas, ammonia and acids (especially hydrofluoric acid), however dilute, must not be used. If any of these cleaning agents is to be used on either the frames or the walls adjacent to the glass, the glass must be protected completely. The use of any proprietory brands of paint stripper on the glass should also be forbidden.
- Whole panels should never be immersed in chemical baths to free the cames from the mortars or putties holding them in place. Skilled conservators can

(a)

(b)

Glass from Kirk Sandall, Yorkshire, before (a) and after (b) successful cleaning. The method used was very gentle and involved minimum intervention. Disfiguring surface dirt has been removed, retaining and revealing essential historical details. The hand is the only portion that has been renewed. (Photographs: Keith Barley) judge from experience whether or not a panel is stable enough in paint or glass to be able to sustain submersion in pure water in order to saturate the bonding agent and achieve the easier stripping of cames.

- Any method of cleaning which scrapes or scratches the exterior surface of the glass is to be avoided. Scratching and scuffing of the surface, however minute, accelerates the development of pitting and surface corrosion wherever the surface tension of the glass has been broken. Mechanical methods such as abrasive blasting, power tools and abrasive pads should never be used.
- Cleaning with air abrasive, ultrasonic, laser or a dental drill should never be attempted by unskilled operators.
- Glass should never be left unprotected if the building facade is to be cleaned.
- The surface of cleaned glass and cames should never be coated with any type of 'protective' varnish or sealant.

How to approach the removal of dirt and corrosion products

The state of the art
Over the years considerable sums of money have been spent on research establishing the nature of the corrosion of historic glass, but little has been spent on finding appropriate methods to clean and conserve it. As a result there exists an extensive bibliography on the mechanisms of glass corrosion and the analysis of the chemical or organic composition of corrosion products but there is virtually no information on how best to remove corrosion products and prevent the process recurring. This places a heavy burden of responsibility on the conservator to use his skill and experience on the bench to determine the most appropriate method to use, and the extent of cleaning to undertake. As with almost all conservation work, the removal of dirt and corrosion products is a question of degree: the degree of the tenacity of the surface accretions, and the degree of effectiveness of progressively interventionist methods deployed to remove it. It should never be forgotten that glass is a two-sided medium, and detailed examination of the nature and extent of the dirt and corrosion on both surfaces is essential before any work commences. Removal of any surface accretions should never damage any paint, enamel, washes or stain. Some surfaces can be easily cleaned *in situ*, others require breaking down into each individual piece of glass followed by detailed work using a microscope to ensure paint lines are not damaged.

An evaluation of cleaning methods
The following list describes several methods which are considered appropriate for cleaning historical glass. The success of each method depends greatly on the technique with which it is applied. The cleaning of historical glass should, therefore, only ever be undertaken by skilled operatives.

1 *Water* is the simplest and most basic cleaning agent. Distilled or deionized water is preferable to any mains water. Hard waters should not be used. Soft cloths or brushes with sprayed water can be used either *in situ* or on the bench to remove loose dirt and bird droppings. This is usually the first cleaning

method to use to reveal the extent of the corrosion concealed beneath, and to enable the conservator to decide which method to use to remove more tenacious dirt and corrosion products. High pressure water jets and steam cleaning are not advised. Soaking or surface saturation should never be used on flaking paint or shaling, decomposing glasses.

2   *Reagents* such as Calgon and EDTA (ethylene diamine tetracetic acid) have been used successfully by skilled conservators to remove both dirt and corrosion without resultant damage. This method is recommended for the individual treatment of pieces of glass either *in situ* or in the conservation workshop. The degree of cleaning for each piece is easily controlled. When the degree of cleansing required is reached, the fluid should always be completely removed with water. Total immersion, poulticing or soaking are not recommended, as the degree of cleaning required for different pieces of glass may vary widely.

3   *Ultrasonic cleaning* is used in many workshops, but only by experienced operators. The length of submersion is critical and should never be more than 6 minutes. The conservator should always establish the chemical composition of the solution used in the ultrasonic bath, and should never use a solution of unknown composition.

4   *Glass fibre brushes* in the hands of a skilled operator working with the aid of a microscope or magnifying lens are the most gentle, most effective and easiest to control of all the dry methods of cleaning and are particularly suitable for *in situ* cleaning. This method is particularly advised for glass which is stained, painted or enamelled.

5   *Dental drills and descalers*, with the assistance of a microscope, are useful tools in the hands of a skilled operator. While they can be tedious to use and difficult to control, they are effective in the excavation of corrosion pits.

6   *Laser beams* are a recent development which has proved successful in removing corrosion products, but they are very expensive and require extensive training of the operative.

7   *Air abrasive pencils* are used extensively for the removal of hard crusts of corrosion products of dirt. There are many variables in controlling the air pressure, length of time of application, angle and distance of application, type and size of nozzle, size and type of abrasive material which must be selected and applied skilfully throughout.

8   *Mechanical grinding* is a last resort, and can only be used on surfaces where there is no paint, enamel or stain. This technique is not generally advised for several reasons. The surface of the glass is completely removed by this method and as it is often difficult to ascertain whether or not paint is present on the surface of the glass beneath a hard crust of corrosion products the paint may also be totally removed. Furthermore, grinding reduces the overall thickness of glass and therefore the intensity of tone of pot metals. It also removes any surface flashing or variations, leaving a disfigured matte surface which may require acid polishing to restore lucency.

The legibility of a design on historically important glass can be acceptably recovered by the reversible method of back plating. On the left-hand side is the new backing plate onto which the missing portions of the book and clasp have been painted. The backing plate has also been moulded to exactly match the contours of the original glass, on the right. The backing plate is glazed on the exterior of the original glass thereby protecting it. Cracks in the original glass are mended before reglazing. Successful back plating requires very highly developed skills. (Photograph: Jill Kerr)

2.8   REPAIR METHODS

Treatment of paint loss

There are only two reversible methods of dealing with paint loss at present; cold paint and back plating.

Cold paint (polymers, acrylics, water or oil-based pigments)
Colour-matched tones applied carefully to the areas of lost painted line, wash, stain or enamel are used extensively in museums particularly to replace lost enamels from medieval glass. Such treatments require constant monitoring as paint often crizzles, crazes and flakes off as a result of the expansion and contraction of the surface of the glass. Reversible, water soluble paints are used to fill in missing portions where it is absolutely certain that the line, or extent of the original design, can be recreated. Because of the volatile nature of cold paint and

its water solubility, it can really only be used effectively in museum conditions, where temperature changes are minimal, there is no threat of condensation, monitoring is easy and the glass can be removed for further care the moment any alteration occurs. This technique should never be used on *in situ* historically important glass.

Back plating

The only acceptable method in use for recovering the legibility of historically important glass. The missing portion of the design or the lost dimension of the colour or stain, are carefully and accurately copied onto a thin piece of modern glass. It is then fired onto the interior surface (in the case of line, stain or enamel), being moulded in the kiln to match exactly the contours and irregularity of the exterior of the original glass. This is essential, as the seal around both pieces of glass when leaded back into the panel has to be complete or condensation can form. Back plates are always signed and dated by the conservator for future reference. The method is generally used to 'restore' lost inscriptions, missing portions of heraldry, important design features, lost flashing or stain and diminished depth of pot metal colours. It is also used, but with great care, where it is possible to ascertain missing backpainting or line and wash detail of faces and hands, or other essential elements of the design. Photographs should be taken of the painted pieces to be plated both with and without plating. Back plating is completely reversible, and, if properly glazed, can be used for *in situ* glazing.

The advantages and disadvantages of back plating are discussed in detail on pp.58–59.

Total replacement

Total replacement of missing pieces should only be undertaken in consultation with a glass historian, who can direct the glass painter to the most stylistically sympathetic examples for copying. In the case of reverse or repeat cartoons, architectural designs, inscriptions or heraldry this is relatively uncontroversial. For heads and attributes it is more problematic. Again, in principle, this method of increasing the legibility of a window is reversible. Proper recording and the retention of the pieces of glass that are removed or superseded (in the case of inept later restorations) are as essential as inscribing the replacement pieces with the artist's name and the date.

Cracks and edge mending

The choice of method can only properly be resolved by the expert conservator on the bench. Some conservators always prefer to use a physical rather than a chemical bond between broken pieces. This can take the form of a thin mending lead, the insertion of a proper came or the use of copper foil.

Mending leads are superficial, and can be soldered into place usually applied to one surface and are used to provide support for breaks or cracks with the glass still *in situ*. They have a limited structural life and always require replacement on releading.

Releading, using a proper came as an integral part of the panel, requires deglazing and the grozing of the broken edges to allow for the increased areas of the inserted lead. This can cause loss of the integrity of the design, and has to be carefully judged aesthetically against the advantage of the increase in strength.

Copper foiling requires expertise and skill. It creates a perceptible line across a break and can therefore impair legibility, but offers minimal interference if backplating and edge joining are inadvisable.

Chemical bonding with glues is another intervention that requires skill and experience. Care should be always be taken to ensure the use of a glue that will not discolour after exposure to sunlight, is not rigid and unresponsive to the movement of glass, can be removed in the event of failure, and which will hold the bond adequately. Self-curing silicone glues have been found to be the most successful bonding agents for edge mending of glass. Where chemical bonding is used, conservators invariably recommend the use of supportive external backing plates, except where the glass is to be presented in museum conditions. No method has been devised to eliminate the 'flash' from the edge of bonded broken glass when viewed from certain angles. As so much disfigurement and damage has resulted from the past use of unsuitable glues in glass conservation, the conservator should always specify the type of silicone glue he is proposing to use in the conservation records. Care should always be taken to make sure that no glue adheres to either surface of the glass beyond the precise area of the edge to be bonded. Clumsy application of glues which stick the backing plate to the exterior surface usually cannot be rectified without extensive damage.

2.9 SPECIAL CONSERVATION PROBLEMS

Plain glazing

Handmade plain glazing is greatly undervalued and is becoming the fastest diminishing of all types of historical glazing. The intrinsic importance of the variations in lucency, colour, form and design of this form of glazing are yet to be fully appreciated.

It is lamentable that plain glazing has usually been replaced with modern machine-made white glass. Modern glass does not replicate the variations in hue, range of tone and light-bending properties of handmade glass. The effect of light passing through original fifteenth-, sixteenth-, seventeenth- or eighteenth-century plain glazing into a historic interior can be greatly altered and subverted by the removal of the original glass. Where original plain glazing survives, it should be retained, preserved, mended and re-used. Some specialist glaziers have extensive stocks of this valuable material, built up from purchasing discarded handmade glass which is being sold off for cullet. It is very important that the texture and type of replacement glazing should be carefully matched. Some European glass manufacturers are reviving the techniques of handmade glass to meet an increasing demand for this special material. It is unfortunate that in Britain this demand does not exist sufficiently for a manufacturer to consider it economically viable to reinstate the handmaking of plain glass.

The interior view of this handmade plain glazing shows some of its special and subtle qualities. These have been interrupted by the introduction of modern textured and coloured glass. The quality of the transmitted light is altered and both the internal and external appearances are devalued. In such situations every effort should be made to retain the variations of hue, range of tone and light-bending properties of the original handmade glass. (Photograph: Jill Kerr)

It is practicable for an experienced glazier to repair and reconstruct the characteristic irregularities of pre-nineteenth-century plain glazing patterns by the simple expedient of taking a lead rubbing before dismantling the glass, transferring the design to a glass sheet, and numbering each piece of glass so that the releading can follow exactly the pattern and position of the original. Any replacement pieces can then be cut exactly to size. Unfortunately, it is easier to make a new window with modern glass than to trouble to retain all the irregularities and character of the original. If the intention is for the lead pattern and the original glass to be retained and reused exactly as before, this must be made clear to all involved in the process of its removal, repair and replacement.

Features to look out for and to record are inscriptions on lead or diamond-cut into glass surfaces, the survival of pieces of horn, and the use of pattern-stamped lead grilles glazed in for ventilation. All these should be preserved and reused.

There is also an increasing tendency to replace nineteenth-century reticulated quarry glazing with plain glazing of modern design. It should be remembered that there were sound practical and structural reasons why diamond quarry glazing was used so ubiquitously. It was a practical and successful solution especially for the transfer of loads. The interaction of the glass and lead of diamond quarry glazing meant the considerable weight of the window could be carried from top to bottom without distortion. Replacement designs based on horizontal cut lines do not have this structural efficiency and can suffer distortion resulting in a zig-zag weakening of the infrastructure, the rapid opening of the lead flanges holding the glass under vertical pressure, and the eventual failure of the window as a watertight entity.

Glazing is an important design feature of both the exterior and the interior of a building. The removal of original plain glazing will not only affect the quality of light to the interior and the appearance of a glazed opening, it can also have a detrimental effect on the external appearance. It need hardly be said that the installation of any of the proprietary pressed glasses, double glazing units or plate-glass with appliqué diamond quarry designs should never be used on historic buildings.

Appliqué glass

Appliqué glass is a twentieth-century technique which combines different coloured plastics and glasses. The design is laid out like a mosaic. The usually unpainted and unfired pieces of coloured material (sherds) are cut and stuck onto a base of plain or etched coloured glass which is then fixed in place with the base glass to the exterior. The glues used are typically epoxy based resins. The appliqué glass technique produces a dramatic interplay of palette, and profoundly variable depths of intense colour resulting from the combinations of multiple layering*. In recent years it has been noticed that a number of appliqué glass constructions have begun to shed sherds. In at least one cathedral, a net to protect the public has had to be erected under the appliqué glass lantern to catch falling fragments. The difficulty appears to have been caused by the failure of the glue to hold the surfaces together under the continuous and demanding stress of expansion and contraction due to the temperature changes which the glass experiences. There are also indications that the epoxy resin suffers ultra-violet degradation. At present there does not appear to be a solution to these problems. Most of the artists who designed these constructions are still practising glaziers, and should be able to say which type of glue was used. The British Society of Master Glass Painters and the British Glass Industry Research Association can help to trace the original glazier and determine the cause of the failure. Both organizations are monitoring and coordinating developments in this unresolved problem area.

Dalles de verre

Dalles de verre is a twentieth-century technique of constructing designs from thick slabs of cast glass set in concrete or epoxy/cement mixes, their surfaces often faceted to catch the light, and their shapes achieved by chipping the slab into form. There have recently been some failures reported where the steel cramps set into the concrete blocks have rusted and caused spalling, damaging the glass and

*The intervals between the pieces of coloured glass are usually grouted with a mixture of cement, silver sand, water and PVA glue.

destroying the integrity of the design. The preservation of designs constructed in this manner is difficult and will involve the treatment of the corroded steel and repair of the concrete or resin matrix.

Epoxy/cement mixes for matrices even when reinforced with glass fibre have also suffered failures, with cracks in the matrix often generated from points in the glass. These cause stress concentrations. It seems that part of the complex cause of failure may be thermal behaviour differences between the glass and the resin/cement matrix. It could be that the matrix would be better without the resin, and it would be wise for the time being to restrict the overall sizes of matrices.

The British Society of Master Glass Painters is monitoring cases of decay in dalles de verre with the British Glass Industry Research Association. The Society can help to contact the original designer.

2.10   EXTERNAL PROTECTION

There are two main reasons for installing external protection to historic glazing:

1   To protect the glass from damage by vandals or breakage from the branches of adjacent trees (see below).
2   To protect conserved or fragile glass from the weather and the atmosphere or to lessen the effects of extreme fluctuations of temperature. (See p.57).

Protection from vandalism and breakage

Some insurance companies, including the Ecclesiastical Insurance Group, insist on exterior protection of important stained glass and specify the degree of protection required for coverage by insurance. There are several types of protection currently in use. The relative merits and disadvantages of these are discussed in the following sections.

All types of external protection must be fitted within the framework of the main light and tracery panels to preserve the overall appearance of the window. They should never be fitted across the whole exterior of a window. All fixings should be non-ferrous and be caulked with lead. They should never be pointed round with mastic or mortar. Percussive drills should never be used to create the holes for fixings. Fixings should be made into joints between masonry units or, at worst, so that one unit adjacent to a joint is affected.

Wire guards

When fixed firmly in place these can deflect missiles but not shotgun pellets.

Unprotected iron wire or wire mesh guards should not be used as even galvanized meshes when cut to shape will corrode at the cut ends. Wire mesh grilles should be made by skilled wire-workers to accurate templates provided by glaziers and galvanized *after* fabrication. They should be fixed by glaziers. A thicker wire frame around each guard is also important.

Advantages

- The exterior of the glass is fully exposed to the cleaning action of rain and wind.

- They are relatively inexpensive.

- They can be cut to templates to match the shapes of window openings and tracery lights.

- They can be removed for cleaning, deglazing, stone repair and replacement.

- They do not involve any adjustment to the infrastructure of the window for fixing.

- If plastic coated mesh is used it can be colour matched to be visually un-obtrusive. Dark colours have been used successfully.

Disadvantages

- Copper and iron wires should not be used, as the corrosion and rust will eventually stain and disfigure the masonry fabric below the window and such stains can be extremely difficult and time consuming to remove.

- They do not deter persistent vandals using heavy missiles or guns.

- They can be bent back or smashed to permit a break-in.

- They reduce the amount of light on the glass.

- They may be clearly visible from the interior and can create a visually disfiguring grid pattern across the design of the glass.

- If they are not properly maintained and fixed, birds can build nests between the glass and the guard.

- They need maintenance and renewal at regular intervals.

Plastic sheeting
Plastics are becoming frequently used. There are currently two types on the market: acrylics (Perspex and Oroglass) and polycarbonates (Lexan).

Advantages

- They are relatively inexpensive and easily obtained.

- They provide greater impact resistance than wire guards and can deflect shotgun pellets if a thick enough sheet is used.

- They provide complete coverage of the glazing area.

- A specialist fitter can cut the plastic to fit the window openings very closely.

- They act as draught excluders and the air cavity increases the thermal insulation.

The effect of installing plastic sheeting as protection to the exterior of a historical glass window can be a visual disaster when viewed from the outside. This must be taken into account when the various options of external protective glazing are considered. (Photograph: Jill Kerr)

- They are relatively light in weight compared to glass.

- They are flexible and can accommodate external ferramenta.

- If properly fixed or set within a fixing frame they can be removed for easy access to the glass or repairs to the stone.

- Polycarbonate sheeting can be subdivided horizontally with lead cames corresponding to saddle bars to give an appropriate external pattern and reduce distorted reflections.

Disadvantages

- All plastic sheeting is highly flammable. As fire is such an effective destroyer of glass, plastics should never be used to protect ancient glass or glass of high quality where the exterior is accessible to vandals.

- All plastics gradually surface-craze and take on a milky appearance and lose lucency through exposure to sunlight and through chemical loss.

- Acrylics are more easily scratched than polycarbonates but both are subject to surface scuffing and wind-etching which is not reversible and which accelerates a loss of translucency in association with the crazing.

- Lichens, moulds and dirt easily accumulate on the surface scratches.

- The sheets need to be removed at regular intervals so that the glass can be washed, insect activity cleaned off (especially spiders' webs) and the ferramenta properly maintained.

- They are often installed without ventilation which leads to damage of the glass fabric and the glass and lead panel behind. Inexpert cutting causes rapid deterioration of the edges and the holes drilled for fixing, with frequent cracking and weakening of the panels.

- Plastic panels must be fixed to allow for expansion and contraction in response to temperature fluctuations, otherwise the sheets can buckle, crack, fissure and even detach themselves from the fixings.

- They must be ventilated top and bottom to allow for the free passage of air. If this is not done condensation will occur and damage the glass and lead, as well as providing a microclimate suitable for algae and lichens to form.

- Although plastic panels do not intrude upon the legibility of the glass from the interior until their increasing opacity begins to reduce the light level, the effect on the exterior elevation of flat, reflective sheets can often be detracting.

- Birds often use the top of plastic sheets to perch or to nest on, and guano etches into the surface.

- If inadequately fixed, the sheets can be removed by leeward suction, possibly damaging the glass and its setting.

Glass sheeting

Glass reinforced by wire mesh has been in use since 1898, when it was developed to meet the demand for safety precautions in skylights, roofing and fire doors. It has the advantage of holding together when damaged or broken by impact or heat, but can be fractured by shot and will star craze. It is made with normal sands which give the glass a green tint. Laminated, toughened, plain, cast and plate glasses have all been used to protect stained glass since the beginning of the century. Laminated glass free from the iron oxide that causes greenishness is obtainable. It is relatively expensive but probably provides the highest quality of glass sheet protection.

Advantages

● The exterior is easy to clean.

● The covering of the opening is complete so it acts as double glazing to improve heat retention.

● Visual intrusion on the interior effect of the design is minimal.

● If completely sealed within the window frame it requires no maintenance apart from cleaning.

● It protects the historical glass from most forms of vandal attack by presenting a defensive layer that can be sacrificed and replaced.

Disadvantages

● All forms of external sheet glass change the appearance of buildings with stained glass windows and this may have a detrimental effect on its aesthetic integrity.

● For fixing, a glazing groove has often to be cut or widened in the stone which may damage the moulding and proportions of the window frame.

● Sheet glass is not easy to remove for the repair, cleaning and maintenance of the historical glass or metal fittings.

● If fractured or broken, the whole sheet has to be replaced.

● It is difficult to examine the condition of the exterior of the glass it protects.

● Overglazing fixed without lead came edgings to accommodate differential thermal movement between the glass and the fixing mortar or masonry may permit water penetration. This in turn may lead to condensation problems, mould growth and the collection of water between the glazing.

● If it is totally sealed into the mullions, moisture may be drawn from within the building through the joints of cames.

Protection of conserved or fragile glass from weather, atmospheric attack and extreme fluctuations of temperature

There are four systems in use by conservators: back plating of individual pieces of glass, isothermal double glazing, external protective glazing and chemical coatings. All of these should be installed by the workshop which cleans and conserves the glass, not by a separate contractor.

Backplating

The process of back plating of individual pieces of glass is discussed on p.49. The advantages and disadvantages of the system are as follows:

Advantages

- The process is easily reversible.

- Where there are missing pieces of the design, the legibility of the whole can be restored by painting and firing the missing portions onto the backing plate.

- If the conserved glass is very thin or fragile its life is extended since it is no longer functioning to keep out the elements, and is not subject to the same degree of stress due to temperature fluctuations.

- If fragments of the original conserved glass are missing, small replacements can be added to complete the whole within the same lead thus preserving the integrity of the design.

- The original cut line can be restored even if the broken piece has been subsequently grozed to take a mending lead.

- Later insertions which are obtrusive or inappropriate can be recorded and removed for replacement by more sympathetic or neutral glass.

- The conservation and repair of single pieces of glass and back plating is often all that is required — for example, roundels, heraldic devices, faces, inscriptions, enamelled glass etc.

- If the original intensity of the colour of the glass has been greatly reduced by corrosion or the loss of flashing or thickness of pot metal, then coloured back plating can restore the lost dimension.

- Yellow stain is almost invariably applied to the exterior surface of glass and is frequently lost or destroyed by weathering or chemical cleaning. This design detail can be replaced on a back plate to the exact dimensions of the original, following the surviving paint line of the interior painted design.

- Missing or destroyed back painting can be replaced on the back plate.

- All such restorations of detail, design or colour can be clearly recorded and all back plates signed and dated. This is usually done on the edge where it is concealed by the lead flange and therefore instantly discernible by subsequent conservators.

- If anything goes wrong with the conserved glass it can be deglazed, examined and rectified, using the backing plate for support.

Disadvantages

- Early twentieth-century conservators using this technique frequently glued the glass to be conserved to the backing plate. This has caused a lot of damage. It is especially difficult to reverse, as in most cases individual recipes of animal glues were used and no records were kept.

- Early twentieth-century conservators used completely flat glass for the backing plate. As medieval glass was never completely flat, the original was often cut up to make it easier to stick to the flat back plate, thus damaging the glass further.

- The back plate must be completely sealed against the historical glass by mastic within the lead. If this is not done, a micro-climate is created which can damage the conserved glass.

- The use of extensive or numerous backing plates increases the overall weight of the panel and it may require additional support from integrated ferramenta or tie bars.

- Inexpertly integrated backing plates with inadequate leads or cement/mastic sealant place the protected glass at risk and must be removed and replaced properly as soon as possible after the deficiency is noticed. Adhesive tape or proprietary glues must never be used on the interior painted surface.

Isothermal double glazing

This system has been in use since the 1950s. The conserved glass is removed from the original fixing frame or glazing groove, which is then glazed completely with modern white glass, cut and leaded to follow the main outlines of the design of the original panel it will protect. The original conserved glass is then replaced on a new interior frame specially constructed for the purpose, fixed within the mullions, and ventilated all round to the interior of the building. This system is designed to eliminate condensation, because the interior and exterior surfaces of the conserved glass are in an isothermal environment, i.e. they are the same temperature within the building, and no longer subject to the extreme fluctuations of temperature created by the exposure to interior and exterior differentials.

Advantages

- If the building is properly ventilated the conserved glass is not subjected to condensation and experiences only minimal disturbances from changes in temperature and humidity.

- This system can help insulate the building from heat loss.

- The condition of both surfaces of the conserved glass can be easily monitored by removing each panel from its fixing within its frame.

- Access for any further conservation, cleaning or exhibition purposes is easy.

- If the method of following the main lines of the design and panel division for the cut lines of the external glazing is used, there is minimal visual disturbance from the interior due to parallax or to the exterior elevation.

- The conserved glass is not replaced into the same atmospheric conditions that may originally have accelerated or caused the advancement of decay and deterioration.

- If minimal conservation has been carried out, the conserved glass is effectively in an environment consistent with a 'holding operation'. This is especially important in cases where there are no present solutions to particular problems such as paint loss resulting from the 'borax' problem or under-firing and where continual exposure to condensation and extremes of temperature places the glass at risk. The isothermal double glazing system of protection may even extend the life of the glass until solutions are developed for problems which cannot be dealt with at present, and at least ensure that there is something left to conserve. This system of 'moth-balling' the problem is usually preferable to experimenting with techniques that have yet to be proved effective, and which are, by and large, not reversible.

Disadvantages

- Isothermal double glazing is the most expensive solution of all the environmental protection layers because of the additional cost of construction of the interior frame to hold the conserved glass.

- In order to eliminate halation around the re-set conserved panels, it is often necessary for a light-excluding border to be glazed around the outer edges of each panel, which increases the cost further, but decreases the vulnerability of the original conserved glass at its weakest point.

- The glazing interspace determines the amount of ventilation between the two surfaces. If it is too small the passage of air is restricted, while if the space is too large, the effect on the proportions of the interior mouldings can be unsatisfactory. It is essential that an architect be fully involved with the design and dimension of the fixings and the frame to avoid these difficulties.

- The exterior layer of isothermal double glazing does not always effectively protect the glass from attack by vandals and it may still require an external guard.

- Condensation can still occur on the conserved surfaces if the building is not properly ventilated.

- For security it is necessary for the conserved glass to be firmly fixed in position.

- If the new glass installed in the exterior glazing groove does not have a leaded design based on the design and panel divisions of the conserved glass, the lead pattern can disfigure the interior effect of the original. If a plain sheet is used, it is the exterior appearance which suffers.

External protective glazing

The conserved glass remains in its original glazing groove and the protective modern glazing is inserted outside it. The interspace is ventilated to the exterior of the building.

Advantages

- The conserved glass does not have to have a specially constructed frame but remains in its original groove.
- There is no halation problem from the edges of the conserved glass.
- The system is cheaper than isothermal glazing as fewer modifications to the existing arrangements are necessary.
- The conserved glass does not have to be weathertight.
- Condensation is reduced.
- The integrity of the interior mouldings and proportions is retained.

Disadvantages

- The whole of the exterior glazing has to be removed to monitor and clean the exterior of the conserved glass unless it is properly hinged.
- The conserved glass is still subjected to exterior atmospheric pollution.
- The additional exterior glazing can create technical and aesthetic problems if the rebate must be widened or part of the exterior moulding lost to allow for fixing the new glazing into the stonework.
- There can still be extremes of temperatures between the interior and the exterior of the conserved glass.

Chemical coatings

It is not yet possible to be confident about the long term performance and hence the protective value of synthetic coatings on historical glass. They are also controversial because they are not reversible. Most synthetic coatings require painstaking and time-consuming application to the exterior surface of each individual piece of cleaned glass before re-leading, and are therefore expensive as well.

In the past, various proprietary varnishes and early plastic colour-enhancing substances were used, often with disastrous results. Cracking, crizzling or shaling off which take the painted lines or surface with it, and severe discoloration are common features of this early period of experimentation. Cellulose nitrate plasticized with camphor (Bakelite) was one of the most damaging coatings used at the beginning of this century. It eventually formed an opaque layer which contracted, removing the surface of the glass and any associated loose paint. As records were rarely kept at this time, any supervening substance on the surface of the glass should be identified and advice sought on its removal, as further experimentation may only increase the damage. The Ancient Monuments Laboratory

The consequences of neglect of a historical glass window can be at best costly to rectify and at worst disastrous in terms of loss of valuable original material. In the instance shown, not only the glazing but the stone traceries have been allowed to deteriorate to the extent that they will soon be no longer functional. The co-ordinating role of the architect in bringing together complementary craft skills and understanding the role of all the glazing and masonry elements and their environment is imperative (Photograph: Jill Kerr)

and the Research, Technical and Advisory Services of English Heritage can advise on where analysis can be carried out and should be consulted in the event of difficulty. Reference 3 contains an extensive analysis and bibliography on the subject of resin coatings (Viacryl), wax coatings, the Jacobi process, hydrophobic and monomolecular layers. The British Glass Industry Research Association can also provide an updated list of recent research papers.

At this time, chemical coatings are best not used at all.

Conclusion
There are no perfect solutions for the complete protection of glass. Each situation and problem is specific and must be individually assessed. By its very nature glass will always be vulnerable to breakage, damage and deterioration. All that can be done is to protect where possible and preserve where feasible, always keeping photographic and written records.

Guidelines and recommendations for the protection of historical glass
The following guidelines are intended to assist those responsible for the protection of historical glass. They are not exclusive to ecclesiastical buildings.

- Assess the advantages and disadvantages of external systems before making a selection. The wrong choice, for the wrong reason, installed by the wrong person can irrevocably damage both the building and the glass which is to be protected.
- Always seek architectural advice before fitting any form of external protection or engaging a specialist glazier.
- Seek the advice of the Diocesan Advisory Committee and the Council for the Care of Churches in relation to ecclesiastical building. Both bodies will gladly advise all religious denominations.
- Ask the local authority conservation officer for names of other owners of historic buildings in the area with similar problems.
- Find out (from any of the above) where local examples of the types of external protection under consideration can be seen, so that the effect on the glass and building can be judged and the practicalities of its installation can be discussed.
- Consult the insurance company on its requirements for acceptable protective covering.
- Have vulnerable and important glass fully recorded photographically. CVMA Archive at the National Monuments Record can advise on this.
- Consult the police if vandalism is persistent for advice on deterrent lighting, restricted access, alarm systems and surveillance.
- Notify the police of the location of particularly valuable glass so that it can be included on a patrol itinerary.
- Seek architectural advice on the treatment or replacement of rusting or corroding guards that are staining the stonework.
- Remove ivy and creeping plants from wire guards or the surfaces of any protective glazing. (See Volume 1, Chapter 2, 'The Control of Organic Growth').

- Ensure that fixings of a protective system cause minimum damage.
- Engage the conservator in a proper system of 'after-care' surveillance and monitoring so that further problems can be detected and rectified before they become serious, damaging and expensive to deal with.
- Include proper maintenance of all types of external protection in a regular programme.
- Keep the external surface of plain or sheet glazing free of dirt, lichens, algae and mould by using an appropriate biocide.
- Monitor plastic sheets for cracking, crazing, the working loose of fittings, buckling and increasing opacity. Inform the architect of any such development.
- Replace broken or failed exterior protection, having sought architectural advice.
- Ensure that at all times exterior protection is properly ventilated and that water collecting at the base will run out freely.
- Make sure that ventilation gaps and holes on any type of protective glazing are not blocked and have grilles fixed to prevent further blockage and access for insects, small animals and birds.
- Remove birds' nests from any system of external protection and refix it properly to ensure that there is not recurrence.
- Do not select a protective system simply because it is cheap or can be done on a do-it-yourself basis. This approach invariably creates more problems than it solves and leaves those involved financially liable to rectify damage.
- Do not hesitate to ask for professional advice on any aspect of exterior protection. It will always cost less than any damage that will ensue from an amateur approach and inadequate coverage.
- Strenuously avoid ready-made commercial double glazing as it is wholly inappropriate for use on a historic building. The proportions of glazing, its materials and such installations are invariably unsatisfactory and damage to the fabric is expensive to rectify.
- It is worth drawing the attention of the fire brigade officer to the locations of important historic glass to avoid its being smashed to provide access for firefighters in the event of a fire.

2.11   SOURCES OF FURTHER INFORMATION

Information on *specialist conservation workshops* can be obtained from The Council for the Care of Churches, 83 London Wall, London EC2 5NA, Tel: (01) 638 0971/2. This is the only national body which keeps a detailed up-to-date list of independent conservators.

The British Corpus Vitrearum Archive at the National Monuments Record of the Royal Commission on Historical Monuments for England, Fortress House, 23 Savile Row, London W1X 2JQ, Tel: (01) 734 6010, is open to the public from 10am to 5.30pm Monday–Friday. The Archive aims to provide comprehensive coverage of the surviving medieval glass in Britain, and holds computerized records of *photographic and related documentary material* on glass of many periods from all contexts. Copies of the conservation records of glass conserved with grant aid from the Historic Buildings and Monuments Commission for England (HBMCE) and the CCC are held here, and the Archive welcomes material to augment the collection. The RCHME can advise on all aspects of *photographing glass* and is particularly concerned to be informed of requests to record threatened glass, damaged glass and glass at risk. Written requests for background on the dates and iconography of glass will be answered if accompanied by clearly legible photographs.

The Glass Information Centre of the Worshipful Company of Glaziers and Painters of Glass, Glaziers Hall, 9 Montague Close, London Bridge, London SE1 9DD, Tel: (01) 403 3300, acts as a clearing house and contact point for information on the *trade and supply of glass* in the UK, and contains the London Stained Glass Repository, run in collaboration with London Division of the HBMCE, which *salvages and secures unwanted stained glass for re-use*. They also provide information on all aspects of *training and apprenticeship schemes* for glaziers.

The International Council of Museums Committee for Conservation of Glass (Co-ordinator, Dr N.H. Tennent), Glasgow Museums and Art Galleries, Kelvingrove, Glasgow G3 8AG, will respond to written requests for information and advice on all aspects of *museum conservation* of glass.

The British Society of Master Glass Painters, 6 Queens Square, London WC1, will respond to written requests for information on *practising glaziers* and all aspects of *commissioning contemporary glass*.

The British Glass Industry Research Association, Northumberland Road, Sheffield, South Yorkshire, S10 2UA, Tel: (0742) 686201, provides a specialized library and information services on all aspects of glass manufacture. It provides a focal point for *technical information* and innovation in all glass-related *research* in both industry and academic centres.

REFERENCES

1 Caviness, Madeline Harrison, *Stained Glass Before 1540, An Annotated Bibliography*, G.K. Hall & Co, 70 Lincoln Street, Boston, Massachusetts, 1983.

   An extremely useful and informative publication on medieval glass. It contains all aspects of glass conservation, techniques of glass painting, historical studies, topographical material, indexes of glass painters and designers, collections, sales and exhibition catalogues, an excellent topographical index, and an author index. The introduction is a particularly lucid and interesting account of the subject.

2 Lee, L., Seddon, G. and Stephens, F., *Stained Glass*, Mitchell Beazley, 1976.

   The best general book on glass as an artist's medium. Includes a brief history of glass, and an account of its manufacture and conservation. This book is superbly illustrated and provides an excellent overview.

3 Newton, R.G., *The Deterioration and Conservation of Painted Glass: A Critical Bibliography*. Published for the British Academy by the Oxford University Press as Corpus Vitrearum Medii Aevi Great Britain, Occasional Papers II, 1982.

   A highly individual and informative book which covers Professor Newton's views on the state of the art and science of glass conservation in 1982.

See also the Technical Bibliography, Chapter 4

Record of conservation
Stained glass

Building
Place   Population   Postal Town   County   Diocese

A: BEFORE WORK

General note
It is strongly recommended that the Corpus Vitrearum Medii Aevi system be used for *numbering* individual windows, especially where more than one window is involved. Wherever possible, please include a sketch plan showing the numbering of the windows.

Report on present condition and proposed treatment
This will normally have been submitted before a case receives a grant. Extra copies of photographs may be required to make up a total of 3 of each (see sheet D). The report should be set out by the conservator on a separate sheet and should include:

Name and address of conservator

Description of glass
Give date, identification of subject, position in building, dimensions, and features of particular interest.

Previous treatment
Give date, practitioner(s) and description of work done if known.

Condition of glass and immediately associated parts of structure
As well as the glass itself, the conservator should also comment on the condition of the ferramenta and immediately associated stonework as well as any other defects which could threaten the glass.

N.B. It is a condition of all applications that a separate report on the condition of the stonework of relevant windows is submitted by the inspecting architect. It is a condition of all grants for stained glass that the architect is commissioned by the parish to be present when the glass is removed and that all stonework repairs necessary in the architect's opinion are carried out at the parish's expense before the glass is returned.

Specification of work required
Give full details of work required (including materials and methods) to each window and detail the cost separately. Estimate separately for the cost of any treatment required to the ferramenta. (Re-arrangement and strengthening of line or area paint work is not generally encouraged unless there are special reasons. Conservators should give full details of any proposals for such work (including right-siding) with their reasons.)

  The conservator should consider whether secondary glazing is required either as protection against environmental pollution or vandalism or both and submit separate estimates for each window as appropriate.

Figure 2.2   Record of conservation (sheet A)
          The Corpus Vitrearum Medii Aevi/Council for the Care of Churches system for recording the condition of and conservation works undertaken on historic glass

Record of conservation

Stained glass

**B: To be filled in by the Inspecting Architect
after completion of work**

Place   Town or village

Condition of building
Name and address of architect in charge

Length of time associated with building

Date of last Quinquennial Inspection

Name and address of the architect if different from above

Give general condition of fabric (mention recent work and current programme). Give note on method of heating. Are the ventilation and rainwater disposal thoroughly satisfactory? Is glazing watertight?

Has work on the glass and any associated stonework repairs been completed to your satisfaction?

Any outstanding conditions which may cause deterioration of the glass or stonework (e.g. situation of church, external hazard, any precarious structural features etc.).

Any significant works or conditions which may in the foreseeable future have an effect upon the object to which this application relates.

Revised 3/86

Figure 2.2   Record of conservation (sheet B)
         CVMA/CCC system

Record of conservation
Stained glass
C: After work

Note Separate forms should be used for each window where treatment differs widely

Grant 19 . . . .

£. . . . .

Description of conservation work

Date Began   Finished   Final Cost

Conservator/Workshop

Description of work carried out

General Note
When compiling a conservation record, the conservator should ask him/herself whether it will enable someone else in a hundred years time to tell precisely what work was carried out.

Please include a conservation diagram marked as shown in the attached key. The diagram may *either* be a *good* photograph of the annotated full size 'after' rubbing (these must be printed large enough to enable them to be read easily) *or* a photocopy of a photograph of the glass after conservation.

Re-leading (State size and type of leading and whether complete or partial)

Cleaning method (Please tick box as appropriate)
1   Ultra sonic bath   ☐
2   Glass fibre brushes   ☐
3   Airbrasive   ☐
4   Manual, with soft cloths   ☐
5   Other — if chemical, state method   ☐

Painting or stippling of inserted pieces (general details including name of painter) — individual panes should be marked on conservation diagram)

Figure 2.2   Record of conservation (sheet C)
         CVMA/CCC system

Record of conservation
Stained glass
D: Photographs

Photographic record

Identity of building should be recorded on the back of each print or transparency frame.

Three prints, preferably black and white, are to be included in each case. In addition, colour prints or transparencies should be included wherever possible. Negatives may be sent if desired, in addition to the above. (All negatives received are passed on to the National Monuments Record.)

1   *Before Conservation*   (date:)

Give subject and position in building in each case and continue overleaf if necessary. Photographs of the exterior surface showing the extent of corrosion, state of leading etc. may be included.

(a)   etc.

2   *During Conservation*   (date:)

Give subject and position in building in each case and continue overleaf if necessary.

(a)   etc.

3   *After Conservation*   (date:)

Give subject and position in building in each case and continue overleaf if necessary.

(a)   etc.

Name(s) of photographer(s)

Type of film   Type of paper

Revised 3/86

Figure 2.2   Record of conservation (sheet D)
CVMA/CCC system

List of abbreviations to be used on stained glass conservation diagrams

C   Cleaned – written record will specify methods used
Ea   Edge joined by adhesive – specify in written record
Eb   Edge joined by copper foil
Es   Edge joined using strap lead
F   Artificial filling
Lc   New linework in cold colour
Lf   New linework in fired colour
Ac   Newly painted area in cold colour
Af   Newly painted area in fired colour
Mg   Modern clear glass newly inserted
Mt   Modern coloured or tinted glass newly inserted

Lc, Lf, Ac or Af may be used in conjunction with Mg or Mt to denote new painted colour on newly inserted glass.

Pi   Plated on the front (inside)
Po   Plated on the back (outside)
Ps   Plated on both sides

If plating is tinted 't' may be added. Lc, Lf, Ac or Af may be used in addition to denote new painted colour or plating glass.

n   May be used with abbreviations for plating or modern glass to indicate stippling
S   Sheet lead placed over modern glass – use after above symbols where appropriate

Please add other symbols if necessary for individual projects

Note   The purpose of the diagram is to provide factual information on work included in the current programme of conservation, rather than an assessment of the date of every piece of glass in the window.

Revised 3/86

Figure 2.2   Record of Conservation: CVMA/CCC List of abbreviations

GROUND PLAN showing window numbering system

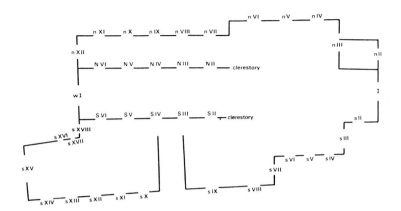

A line is drawn from (liturgical) east to west. The east window is always I, the west w.I. Lower case n(orth) and s(outh) are used for all the windows of the relevant orientation above and below the centre line. Upper case is used for all upper (clerestory) windows. Roman numerals are used to number each opening from east to west whether or not the window contains historically important glass.

WINDOW PLAN and panel numbering system

Numbering always follows the same sequence from bottom to top, left to right. In order to distinguish between main light panels and tracery lights, the number precedes the letter for main lights, the letter comes before the number for the tracery. Both follow a sequential grid pattern directly related to the panel divisions, number of tracery openings and architectural divides. Transom divisons are treated as integral with the panel numberings.

The CVMA numbering system is mnemonic, simple to use and internationally recognized. It obviates the necessity of using lengthy worded descriptions to locate a panel precisely within a building. Although primarily designed for churches, it is based on compass orientation and can be adapted for secular structures. Should you require assistance in using the system, the CVMA Archivist at the NMR (address on p.64) will always answer queries. In many cases a building containing historically important window glass will already have been numbered for the Archive.

Figure 2.3 The CVMA/CCC system for numbering windows and glazing panels
(Jill Kerr)

3 THE USE OF RESIN (POLYMER) PRODUCTS IN THE REPAIR AND CONSERVATION OF BUILDINGS

Edmund King

3.1 THE IMPORTANCE OF KNOWING ABOUT POLYMERS

Problems associated with the use of polymers in remedial work in the building industry, especially with concrete, are well known. These have spawned a wide range of new products based on polymeric materials and described by their chemical name or type. The description of such products rarely gives much clue as to their properties and uses, and it has therefore been necessary for the specifier to become familiar with a new terminology and the properties of such products.

Increasingly, these same products and the contractors who use them are entering the area of conservation, giving architects and engineers the same problem as their colleagues in the field of construction details and concrete repairs. In the context of building conservation, however, the problems are increased, since an assessment has to be made of whether products formulated for these other uses should be used on stone and timber. Frequently they are unsuitable without modification.

Although a detailed chemical knowledge is not essential, it is becoming increasingly necessary that the conservation specifier is conversant with the chemical terminology. It is important that the chemical name can be linked to a product type of known properties and that it can be determined whether these properties are appropriate to a particular problem. This chapter describes briefly some of the product types likely to be of interest in the repair and conservation of buildings and some of their advantages and disadvantages.

3.2 POLYMER TYPES

The polymer types are considered in two groups:

Group 1: Non-structural or emulsion polymers
- styrene/butadiene copolymers,
- styrene/acrylic copolymers,
- acrylic copolymers.

These materials are delivered to site ready for use.

Group 2: Structural polymers
- epoxy resins,
- polyester resins,
- polyurethane resins.

These products are delivered to site in two or more containers and have to be mixed prior to use. Occasionally the polyurethane resins are single-pack systems relying on water or moisture to effect the cure.

Numerous other resin types, such as silicones, alkoxysilanes, and other surface treatments are not described here. (See Volume 1, Chapter 9 'Masonry Consolidants' and Chapter 10 'Colourless Water-Repellent Treatments.) The various applications of group 1 and group 2 polymer types in building conservation are described below in the context of adhesives, mortars, grouts, crack injection, anchors and timber repairs.

3.3 THE USE OF POLYMERS

Adhesives
Both group 1 and group 2 polymers, under the appropriate conditions, are excellent adhesives.

The group 1 adhesives, although different chemically, have similar adhesive properties. They are good adhesives for stone and brick but there is less application for timber. The group 1 adhesives were developed to overcome the water-sensitivity of the PVA (polyvinyl acetate) resins. Having a high resistance to water, these adhesives are suitable for internal and external applications. Being non-structural, they are not suitable for filling voids, although they do form the basis for many cementitious or PFA-based grouts.

The group 2 resins are structural adhesives and can be used to fill voids. However, both the epoxy and polyester resins used for structural repairs are very hard, strong, inflexible materials, and in soft mortars or soft stonework their use is not recommended. Two further disadvantages of polyester resins are that they shrink significantly on curing and, in the longer term, unless stated otherwise, the adhesion to wet surfaces is poor. This contrasts with epoxy resins, many of which have been formulated for excellent bond to wet surfaces. The shrinkage on curing is negligible, providing that the right mix is used. They do have relatively high coefficients of thermal movement. In thin layers such as cracks this is inconsequential, but in filling voids of over, say 15 mm, damage could certainly be caused if neat resin had been used. In large voids neat resin would not be employed but mixed with suitable fillers. Polyurethane resins tend to be more flexible, but to date they are used far less than either epoxy or polyester resins. Although less readily available, it is likely that the polyurethane resins are potentially more useful than either the epoxies or polyesters for adhesion to stonework.

The strength of adhesion of two similar surfaces depends on three factors:

- the cohesive strength of the substrate(s);
- the cohesive strength of the adhesive; and
- the adhesive strength of the adhesive to the substrate.

The strength of the final bond will never be any better than the weakest of these three bonds. If the substrate is weak then the bond will be weak. Therefore it is often necessary to strengthen or consolidate the surface prior to or even at the same time as sticking something to it. Many of the adhesives mentioned above can be used in a diluted form to consolidate a weak, porous surface.

Adhesion to timber can be difficult, although the conservator sometimes has the advantage that old timbers may have a lower moisture and resin content, both of which facilitate adhesion. Many existing woodworking glues are based on urea-formaldehyde or urea-resorcinol resins, and these continue to be used widely. There are also some woodworking glues based on group 1, non-structural, polymers.

The structural adhesives form strong bonds with timber. The most commonly used adhesive is an epoxy resin; usually timber has sufficient strength both parallel to and across the grain to permit the use of strong adhesives. For structural applications the use of adhesive is supplemented with mechanical fixing using stainless steel rods. This will increase the total strength and the flexural strength.

Bonding

'Bonding' in this context is taken to mean the sticking of a repair mortar to a solid substrate. Bonding and adhesion are closely allied subjects. To ensure good bonding the following points should be noted:

1 Chemical cleanliness of work surfaces is imperative (dust and grease free).
2 Different resin types must not be mixed. For a copolymer repair mortar a copolymer slurry must be used and for an epoxy mortar an epoxy resin.
3 The repair mortar must be applied while the bond coat is still 'tacky'. It is suggested in some literature that if the bond coat has cured and is not tacky then a second bond coat should be applied and the repair continued. This is not good advice as the bond between the two coats of bonding agent will be weak. If the first bond coat has cured, the original surface must be uncovered and the process started again.

4 It should be noted that there may be a particular difficulty in wetting stones with resin, especially stones such as basalt, granite or quartzite. The preference of high surface energy materials such as these for water over resin may mean that adhesion is disrupted by resin or stones absorbing atmosphere moisture. Cycling due to applied loads or thermal or moisture movements, especially in combination with wetting agents such as soapy water will assist in the propagation of cracks at the resin/mineral interface.

Mortars

The mortars referred to in this chapter are those which may be used to repair stone and timber, not the conventional mortars used for bedding and repointing.

All the group 1 and group 2 polymers, with the exception of polyurethanes, are used quite commonly in repair mortars with varying degrees of success.

Polymer-modified mortars

All of the group 1 polymers can be used as additives in cementitious mortars. For repair mortars the polymer additives are usually formulated to increase the tensile and flexural strengths while reducing or keeping the same compressive strength. Other advantages cited for polymer-modified mortars include the production of a waterproof mortar, improved workability and the ability to use the mortar in thin applications. The improvement in waterproofing properties is well established. However, the other claims need clarification. In skilled hands there is an improvement in workability, but it is possible to over-trowel, giving a polymer-rich and unnatural surface. The successful use in thin applications is not proven and it is not recommended that traditional thick coat renders should be repaired or replaced with a thin coat polymer-modified mortar/render.

The role of lime in traditional sand:lime and more recent cement:sand:lime mortars was more than that of a plasticizer. While the addition of a group 1 polymer or an air entrainer to a 3:1 sand:cement mortar will make it easier to work and place, it will not provide the cured mortar with the ability to accommodate dimensional change, as would lime.

Resin mortars

Epoxy or polyester resins form the basis for most resin mortars. The resins are filled with between 80 and 90 per cent of special dry graded sands; the final products, when cured, give compressive strengths around $80-100$ N/mm^2, although this can be reduced to around 40 N/mm^2 by incorporating lightweight fillers.

Except, very occasionally, for the toughest stone or brick, the use of resin mortars should be discouraged in masonry.

The strength of resin mortar is related closely to the strength of the filler, provided high filler loadings are used, i.e. over 85 per cent. It is possible, for example, to substitute dry, crushed stone dust for sand and obtain a repair mortar approximating in strength to the original stone with an acceptable permeability. Stone dust filler can be used in a 19:1 ratio with resin. Hand mixing of such materials is virtually impossible. The right type of mechanical mixer is required for thorough integration. A resin-aggregate system formulated on site is unlikely to have good trowelling properties; over-trowelling will give a resin-rich surface. The method may be suitable for small patch repairs, but for larger areas a factory-formulated product is advisable. These kinds of repair are only suitable in some special and local situations and must never be used in large areas to overcome surface failures (see Volume 1, Chapter 4 'Repair with Mortar ('Plastic Repair')').

Grouts

Grouting, in this context, is the filling of cracks and voids to improve the structural stability of an area of masonry and/or to reduce its permeability to water. Grouting equipment is described in Volume 1, Chapter 3, 'Grouting Masonry Walls'. Equipment is still being developed for the controlled placing of grouts which will be particularly suitable for vulnerable historic masonry.

Cementitious/PFA grouts

These grouts are mixes of cement and/or PFA (pulverized fly ash) with various additives, such as bentonite (or other clays) to prevent segregation and reduce shrinkage, and one of the group 1 type polymers to improve the waterproofing and flow properties. It is this class of grouts in particular for which a controlled method of application is necessary.

Chemical grouts

As with chemical mortars, the most common chemical grouts are epoxy resins with suitable sand fillers. The particle size distribution of the filler is different from that of mortars, and to achieve flowability the filler to resin ratio is appreciably lower, i.e. around 3−4:1 compared to 6−9:1 for a mortar, these ratios being by weight for normal sand fillers.

Because of the higher resin content in a chemical grout, the movement characteristics of the product under constant loading, i.e. creep, must be considered. Prior to using any chemical grout system it is advisable to obtain creep data relevant to that particular product from the manufacturer. Because of the scarcity of such data, this request will eliminate most of the chemical grouts on the market.

A potentially interesting chemical grout is a urethane system which is premixed with a catalyst on site; on contact with water or moisture it reacts to give a closed cell polyurethane foam. The increase in volume on foaming fills the voids and as the foam is closed-cell the grout is impervious to moisture. Recent field trials have shown that the method of application of this grout needs to be refined before a full evaluation can be carried out.

All the chemical grouts mentioned fill voids. Assuming the application problems can be solved, the lack of creep data and the very high strength of the chemical grouts would point towards the cementitious grouts usually being preferred. Comments on the polyurethane foam system must await further evaluation, but in many situations it is useful to have a grouting system which adds little extra weight to the existing structure.

As voids in buildings are frequently damp, the use of polyester-based products should be avoided because of bonding problems (see 'Adhesives', above).

Crack injection

The filling of cracks in the brickwork and stonework of old buildings is becoming increasingly popular, presumably due to the success of sealing cracks in concrete. However, there are significant differences which must be recognized. A common cause of cracking in concrete is the corrosion of the underlying reinforcement, and sealing such cracks can arrest the corrosion process if water and air can no longer get to the steel. This is unlikely to be the case in historic masonry. Many modern building materials tend to be hard and stiff compared to the older materials, which are softer and more elastic and better able to accommodate movement.

Nevertheless, cracking does occur in historic masonry and these cracks can be static (e.g. settlement), progressively widening (e.g. subsidence) or widening and narrowing in a cyclical fashion (e.g. thermal movements). Prior to carrying out any crack repair, the cause should be ascertained and, if of the first two types, remedied. Even when this is done the crack may continue to move due to thermal changes in the fabric. In some situations it will be preferable to inject the resin such that, once injected and cured, it remains mostly in compression. For narrow cracks one of the group 1 polymers or an epoxy resin can be chosen, although it is preferable to use one of the group 1 polymers. For wider cracks it may be necessary to use the gap filling properties of the epoxy resins, although polyurethane resin has been used to seal cracks successfully. The expansion risks of epoxy resin mentioned earlier must be taken into consideration.

Of all the resin repair methods advocated for the repair of old buildings, it is probably the area of crack injection which has been most oversold and most abused. Resins have a valuable but limited use. They are not the universal answer to repairing cracks.

Resin anchors

Resin anchors or 'stitches' are used to strengthen and consolidate stone and brick structures or timber beams or to tie together, where required, different parts of a building. The resins used are epoxy or polyester resins. These are used to 'grout in' stainless steel or polyester dowels to pre-drilled holes, preferably having used a vacuum drill. The length, diameter and position of the rods is determined on site.

Timber repairs

Over the last twenty-five years epoxy mortars have been used quite extensively and quite successfully to repair timber beam ends. Combined with the use of stainless steel or polyester rods, the epoxy mortars have sufficient strength to perform the same role as the original timber. Repairs can be carried out without disturbing other parts of the building and, in certain circumstances, this can offer real advantages.

The choice of this method compared to 'splicing in' a new piece of timber depends on several factors including accessibility and the condition of various elements of the building. If a resin grout is suggested rather than a resin mortar, the problem of creep should be considered if a permanent force is to be applied. The use of epoxy mortar repairs alone along the length of the beam is debatable. If the beam needs strengthening, and traditional timber repairs are not possible, the best approach could be to grout in stainless steel or polyester rod dowels with an epoxy or polyester resin. (See p.19 section 1.11)

3.4   SITE RISKS

Every possible precaution must be taken to protect surfaces against runs and spillages of resin. Escaping resin can be swabbed with solvents such as xylene, toluene or benzene, but it must be remembered that these are unpleasant and hazardous materials which should only be used in well ventilated conditions. Dimethylformamide (DMF) is better and, being fully miscible with water, can be washed and flushed away. Unfortunately most solvents will cause a darkening of masonry surfaces which will not readily disappear. Cured resin can most satisfactorily be removed by mechanical means such as a small, sharp chisel. If air abrasive tools, miniature discs or compressed air chisels are used, which are liable to create dust (especially particles of 5 microns or less), then masks fitted with good dust filters must be worn.

All resins and all solvents should be treated with respect. The dermatitic effect of hardeners and the severe de-greasing effect of solvents make the use of gloves imperative. Barrier creams are also strongly recommended.

The use of resins in complicated situations or where large quantities may be needed should be limited to expert handling. Unfamiliarity with materials, techniques and hazards can create problems which are not easy, or may be impossible, to put right.

3.5   CONCLUSION – THE NEED FOR CAUTION

This chapter provides a summary of some of the newer polymeric materials, their uses and potential misuses in the repair of old buildings. Caution is advocated in extrapolating from the results achieved with the same polymers with concrete to the results likely to be obtained on stone, brick and timber.

Resin and resin techniques should not be thought of as quick answers to building repair problems or as a substitute for traditional materials and methods. The use of these relatively new products in the context of historic building fabric requires a high level of skill and experience. Few, if any, resin repairs are practicably reversible, and the decision to use them and the way in which they are used should be made and carried out only by those with specialist knowledge and trade skills. Properly specified and applied, however, some polymeric materials can play a vital role in building conservation.

Further Information
At this time the best sources of more detailed and technical information on the properties or resins are the manufacturers of the structural, non-structural, polyester and polyurethane resins, rather than their formulators or blenders.

4 SELECT TECHNICAL BIBLIOGRAPHY

This bibliography is a selection of useful references relating to practical aspects of the repair, maintenance and conservation of historic buildings.

4.1 CONSERVATION AND BUILDING SCIENCE

Addleson, Lyall, *Building Failures*, The Architectural Press, London, 1973.

APT Bulletin — Index, Association for Preservation Technology, Ottawa.

Ashurst, John and Malnic, Nicola, 'Options to Bodging,' *AJ*, Special Renovation Issue, 21 & 28 August 1985, pp 40–66. (Deals with stone, brick, terracotta and render.)

Bowyer, Jack (ed), *Handbook of Building Crafts in Conservation*, (Commentary on Nicholson's *New Practical Builder and Workman's Companion*, published in 1823) Hutchinson, Leeds 1981.

British Standards Institution, BS 3589 *Glossary of General Building Terms*.

Chambers, J Henry, *Cyclical Maintenance of Historic Buildings*, US Government Printing, 1976.

Clydesdale, A, *Chemicals In Conservation*, Conservation Science, Kelvingrove Art Gallery & Museum, Glasgow, 1982.

Construction Industry Research Information Association, *A Guide to the Safe Use of Chemicals in Construction*, CIRIA, 6 Storey's Gate, London, SW1P 3AU, 1981.

Cowan, Henry, *Dictionary of Architectural Science*, Applied Science Publishers Ltd, London, 1973.

Crafts Council:
  Science for Conservators, Book 1: An Introduction to Materials, 1982;
  Science for Conservators, Book 2: Cleaning, 1983;
  Science for Conservators, Book 3: Adhesives and Coatings, 1983;
  Crafts Council Conservation Science Teaching Series, London.

Diamant, R M E, *The Chemistry of Building Materials*, Business Books Limited, London, 1970.

Edinburgh New Town Conservation Committee, *The Care and Conservation of Georgian Houses*, Edinburgh, 1979.

Everett, Alan, *Mitchell's Building Construction — Materials*, B T Batsford Ltd, London, 1970.

Feilden, Bernard M., *Conservation of Historic Buildings*, Butterworth and Co. Ltd, London, 1982.

Fidler, John, 'Non-destructive Surveying Techniques for the Analysis of Historic Buildings', *ASCHB Transactions*, Vol 5, 1980, pp 3–10.

Hollis, Malcolm, *Surveying Buildings*, Surveyors Publications, London, 1983, 2nd ed, 1986.

Holmström, Ingmar and Sandström, Christina, *Maintenance of Old Buildings — Preservation from the Technical and Antiquarian Standpoint*, National Swedish Institute for Building Research (from Central Office of National Antiquities, Utvecklingssektionen, Box 5404, 1–114 84 Stockholm, Sweden).

Hughes, Philip, *The Need for Old Buildings To 'Breathe'*, SPAB Information Sheet 4, Society for the Protection of Ancient Buildings, London, 1986.

ICCROM and CAL, *International Index of Conservation Research — Repertoire International de la Recherche en Conservation*, ICCROM and Conservation Analytical Laboratory, Smithsonian Institution, ICCROM, Rome, 1988.

Insall, D.W., *The Care of Old Buildings Today*, The Architectural Press, London, 1973.

Lucas, Clive, *Conservation and Restoration of Buildings: Preservation of Masonry Walls*, Australian Council of National Trusts, Canberra, 1982.

McCann, Michael, *Artist Beware — The Hazards and Precautions in Working with Art and Craft Materials*, Woodson-Guptill Publications, New York, 1979.

Melville, Ian and Gordon, Ian A, *The Repair and Maintenance of Houses*, The Estates Gazette Limited, (151 Wardour Street) London, 1973.

Mills, Edward D, *Building Maintenance and Preservation*, Butterworths, London, 1980.

Parnell, Alan and Ashford, David H, *Fire Safety in Historic Buildings, Part 1 — Fire Dangers and Fire Precautions*, SPAB and the Fire Protection Association, Technical Pamphlet 6.

Powys, A R, *Repair of Ancient Buildings*, reissued by SPAB with additional notes in 1981.

Richardson, Barry A, *Remedial Treatment of Buildings*, The Construction Press, Lancaster, 1980.

Tate, J O, Tennant, N H and Townsends, JH (eds), *Resins in Conservation*, Proceedings of the Symposium held at the University of Edinburgh, 21–22 May 1982, Scottish Society for Conservation and Restoration, Edinburgh, 1983.

Timms, Sharon (ed), *Preservation and Conservation: Principles and Practices*. Proceedings of the Northern American International Regional Conference, Williamsburg, Virginia, 10–16 September 1972, The Preservation Press, National Trust for Historic Preservation in the US, 1976.

Torraca, Giorgio, 'Brick, Adobe, Stone and Architectural Ceramics: Deterioration Processes and Conservation Practices', *Preservation and Conservation: Principles and Practices*, Sharon Timms (ed), The Preservation Press, National Trust for Historic Preservation in the US, 1976.

Torraca, Giorgio, *Porous Building Materials: Materials Science for Architectural Conservation*, 2nd ed., ICCROM, Rome, 1982.

Torraca, Giorgio, *Solubility and Solvents For Conservation Problems*, 3rd ed., ICCROM, Rome, 1984.

4.2   HISTORY OF CONSTRUCTION

Adams, Henry, *Cassell's Building Construction*, Cassell and Co Ltd, London, 1912.

Bowyer, Jack, *History of Building*, Orion, London, 1987.

Briggs, Martin S, *A Short History of the Building Crafts*, The Clarendon Press, Oxford, 1925.

Clifton-Taylor, Alec, *The Pattern of English Building*, B T Batsford Ltd, London, 1962; Faber & Faber 1972.

Clifton-Taylor, Alec, *Cathedrals of England*, Thames and Hudson, London, 1967.

Clifton-Taylor, Alec, *English Parish Churches*, Thames and Hudson, London, 1967.

Clifton-Taylor, Alec and Ireson, A S, *English Stone Building*, Victor Gollancz, London, 1983.

Cruikshank, D and Wyld, P, *London: The Art of Georgian Building*, Architectural Press, London, 1975.

Davey, N, *A History of Building Materials*, Phoenix House, 1961.

Dixon, Roger and Methensius, Stefan, *Victorian Architecture*, Thames and Hudson World of Art Library, 1978.

Everett, Alan, *Mitchell's Building Construction − Materials*, B T Batsford Ltd, London, 1970.

Fitzmaurice, R, *Principles of Modern Building: Walls, Partitions and Chimneys*, HMSO (for the Building Research Station), 1938.

Gwilt, Joseph, *Encyclopedia of Architecture*, Longman, London, 1842, 1876 and 1888.

Knoop, D and Jones, G P, *The Mediaeval Mason*, Manchester University Press, 1933; 3rd ed, 1967.

McKee, Harley J, *Introduction to Early American Masonry*, Preservation Press (US), 1973.

Melville, Ian and Gordon, Ian A, *The Repair and Maintenance of Houses*, The Estates Gazette Ltd, London, 1973.

Middleton, G A T, *Building Materials − Their Nature, Properties and Manufacture*, B T Batsford, London, 1905.

Pain, William, *The Builder's Companion and Workman's General Assistant*, London, 1762, reprinted Gregg International, Richmond, Surrey, 1972.

Prizeman, John, *Your House − The Outside View*, Hutchinson, London, 1975.

Rivington's Series of Notes on Building Construction, *Notes on Building Construction Part 3, Materials*, Longman Green & Co., London, 1892.

Salzman, L F, *Building in England Down to 1540*, Oxford University Press, 1952.

Service, Alistair, *Edwardian Architecture*, Thames and Hudson, 1977.

Singer, Charles, Holmyard, E J, Hall, A R and Williams, Trevor I, *A History of Technology*:
Volume 1, From Early Times to the Fall of the Ancient Empires, 1954;

    Volume 2, Mediterranean Civilisations and The Middle Ages, 1956, corrected 1957;

    Volume 3, From the Renaissance to the Industrial Revolution, c.1500–c.1750, 1957, corrected 1964;

    Volume 4, The Industrial Revolution, c.1750–c.1850, 1958, corrected 1965;

    Volume 5, The Late Nineteenth Century, c.1850–c.1900, 1958, corrected 1965; Clarendon Press, Oxford.

Smith, John F., *A Critical Bibliography of Building Conservation*, Mansell, London, 1978.

Summerson, Sir John, *Georgian London*, Penguin, London, 1945, 1969.

The Builder's Dictionary, London, 1734, republished, Association for Preservation Technology, Washington, 1981.

4.3 BRICK

Avoncroft Museum of Buildings, *Bricks and Brickmaking*, Stoke Heath, Bromsgrove, Worcs, 1978.

Bedfordshire County Council & Royal Commission on Historic Monuments (England), *Survey of Bedfordshire — Brickmaking, A History and Gazetteer*, County Hall, Bedford MK42 9AP, UK, 1979.

Bidwell, T G, *The Conservation of Brick Buildings — The Repair, Alteration and Restoration of Old Brickwork*, Brick Development Association, London, 1977.

Brick Development Association, *Building Note 2: Cleaning of Brickwork*, BDA, Windsor, England.

Bricks in Hertfordshire, Hertfordshire Conservation File, Hertfordshire County Council, 1983.

British Standards Institution:

    BS 3921: *Clay Bricks and Blocks*, 1974;

    BS 6270: Part 1, *Code of Practice for Cleaning and Surface Repair of Buildings — Natural Stone, Cast Stone and Clay and Calcium Silicate Brick Masonry*, 1982;

    BS 187: *Calcium Silicate Bricks*, 1969.

Brunskill, Ronald and Clifton-Taylor, Alec, *English Brickwork*, Ward Lock, 1977.

Building Research Establishment:

    Digest 164: *Clay Brickwork 1*, 1974;

    Digest 165: *Clay Brickwork 2*, 1974;

    Digest 200: *Repairing Brickwork*, 1977;

    Digest 257: *Installation of Wall Ties In Existing Construction*, 1985;

    Digest 139: *Control of Lichens, Moulds and Similar Growths*, 1982;

    BRE, Watford, England.

Clifton-Taylor, Alec, *The Pattern of English Building*, Batsford, 1962; Faber and Faber, 1972.

Collier, Richard, *Guidelines for Restoring Brick Masonry*, The British Columbia Heritage Trust, Technical Paper Series, Victoria BC, 1981.

Davey, N, *A History of Building Materials*, Phoenix House, 1961.

Dobson, E, *A Rudimentary Treatise on the Manufacture of Bricks and Tiles*, 2 vols, London, 1850; reprinted, ed F Celoria, as *Journal of Ceramic History*, whole No.5, with original pagination, vol. 1, pp 39–41; vol 2, p 95.

Earnes, Elizabeth S, *Medieval Tiles: A Handbook*, The British Museum, 1968.

Firman, R J and Eleanor, P, 'A Geological Approach to the Study of Medieval Bricks', *The Mercian Geologist*, vol 2, No 3, December 1967, pp 299–318.

Hammond, Martin, *Bricks and Brickmaking*, Shire Album 75, Shire Publications Ltd, Aylesbury, England, 1981.

Hammond, M D P, 'Brick Kilns: An Illustrated History', *Industrial Archaeology Review*, vol 1, no 2, Spring 1977, pp 171–92.

Harley, L S, 'A Typology of Brick', *Journal of the British Archaeological Association*, 3rd Series, vol XXXVII, 1974, pp 63–87.

Harrison, William H, 'Conditions for Sulfate Attack on Brickwork', *Chemistry and Industry*, 18, September 1981, pp 636–9.

Hillier, Richard, *The Clay That Burns – A History of the Fletton Brick Industry*, available from The Public Relations Officer, London Brick Company, Stewartby, Bedfordshire, 1981.

Lloyd, Nathaniel, *A History of English Brickwork*, H G Montgomery, 1925, reprinted by Antique Book Collectors Club, 1983. (Covers up to 1800.)

McGrath, Thomas L, 'Notes on the Manufacture of Hand Made Bricks', *APT Bulletin*, vol 11, no 3, 1979, pp 88–95.

Mack, Robert C and Look, David W, *Repointing Mortar Joints in Historic Brick Buildings*, Preservation Brief No. 2, Technical Presentation Services Division, Heritage Conservation and Recreation Service, US Department of the Interior, US Government Printing Office, 1980.

Perks, Richard Hugh, *George Bargebrick Esq*, Meresborough Books, (7 Station Road, Rainham, Gillingham, Kent, ME8 7RS) 1981.

Ritchie, T, *On Using Old Bricks in New Buildings*, CBD Series 138, National Research Council, Ottawa, 1971.

Robinson, Gilbert C, 'Characterization of Bricks and Their Resistance to Deterioration Mechanism', in preprints of *Conservation of Historic Stone Buildings and Monuments*, Conference conducted by National Research Council, National Academy of Science, Washington DC 2–4 February, 1981.

Salzman, L F, *Building in England down to 1540*, Oxford University Press, 1952.

Twist, Sidney James, *Stock Bricks of the Swale*, Sittingbourne Society, The Fleur de Lis Heritage Centre, Preston Street, Faversham, Kent, ME13 8NE, 1984.

Volz, John R, 'Brick Bibliography', *APT Bulletin*, vol VII, no 4, 1975, pp 38–49.

Wight, Jane A, *Brick Building In England from The Middle Ages to 1550*, John Baker, London, 1972.

Williams, G B A, *Pointing Stone and Brick Walling*, The Society for the Protection of Ancient Buildings, Technical Pamphlets, reprinted 1983.

Woodforde, John, *Bricks to Build A House*, Routledge and Kegan Paul, London, 1976 (available from The Public Relations Officer, London Brick Company, Stewartby, Bedfordshire).

Mathematical tiles

Buckham, David, 'Mathematical Tiles', Unpublished Dissertation Diploma in Building Conservation, Architectural Association, London, 1979.

Dobson, E, *A Rudimentary Treatise on the Manufacture of Bricks and Tiles*, 2 vols., London, 1850; reprinted ed F Celoria, as *Journal of Ceramic History*, 5, 1971.

Exwood, M (ed.) *Mathematical Tiles: Notes of Ewell Symposium*, 14 November 1981, Ewell, 1981.

Exwood, M, 'Mathematical Tiles, Great Houses and Great Architects', in Exwood, 1981, pp 26–30.

Exwood, M, 'Mathematical Tiles: A Georgian Masquerade', *Period Home*, vol 3, no 6, April/May 1983, pp 28–33.

Loudon, J C, *Encyclopaedia of Cottage, Farm and Villa Architects*, London, 1833.

Powell, C G, *An Economic History of the British Building Industry 1815–1979*, London and New York, 1980.

Smith, T P, *Mathematical Tiles in the Faversham Area*, Faversham Society, 1984.

Tarn, J N, *Working-Class Housing in 19th-Century Britain*, Architectural Association Paper No 7, London, 1971.

Willmott, F G, *Bricks and Brickies*, Rainham, 1972, reprinted 1977.

4.4   STONE

Amorosso, G G and Fassina, V, *Stone Decay and Conservation – Atmospheric Pollution, Cleaning, Consolidation and Protection*, Materials Science Monograph 11, Elsevier, Amsterdam, 1983.

Arkell, W J, *Oxford Stone*, Faber, 1947.

Ashurst, J and Clarke, B L, *Stone Preservation Experiment*, Building Research Station Report, Building Research Establishment, 1972.

Ashurst, J and Dimes, F G, *Stone in Building – Its Use and Potential Today* Architectural Press, London, 1977; (reprinted by the Stone Federation, 1984.

Ashurst, J, Dimes, F G and Honeyborne, D B, *The Conservation of Building and Decorative Stone*, Butterworths Scientific, London, 1988.

Association for Preservation Technology, *Masonry Conservation and Cleaning*, Proceedings of the Conference, Toronto, Canada, 16–19 September, 1984.

Beall, Christine, *Masonry Design and Detailing*, Prentice Hall (US), 1984.

Bowley, M J, *Desalination of Stone: A Case Study*, BRE Current Paper Series CP 46/75, BRE, Watford, 1975.

British Standards Institution:
    BS 2847: 1951, *Glossary of Terms for Stone Used in Buildings*
    BS 3826: 1969, *Silicone Based Water Repellents For Masonry*
    BS 6477: 1984, *Water Repellent Treatments For Masonry Surfaces*
    BS 5390: 1976, (1984) *Code of Practice for Stone Masonry*
    BS 6270: Part 1: 1982, *Code of Practice for Cleaning and Surface Repairs of Buildings, Part 1, Natural Stone, Cast Stone and Clay and Calcium Silicate Brick Masonry*
    BS 1217: *Cast Stone*

Building Research Establishment:
    Digest 125 *Colourless Treatments for Masonry, 1971*
    Digest 139 *Control of Lichens, Moulds and Similar Growths, 1982*
    Digest 177 *Decay and Conservation of Stone Masonry, 1975*

Digest 269 *The Selection of Natural Building Stone, 1983*
Digest 280 Cleaning External Surfaces of Buildings 1985
BRE, Watford, England.

Building Research Establishment,
Technical Information Leaflet,
TIL 64 Damage Caused by Masonry or Mortar Bees,
Building Research Establishment, Watford, England.

Caroe, A D R and Caroe, M B, *Stonework: Maintenance and Surface Repair,* Council for the Care of Churches, London, 1984.

Clifton, James R, *Stone Consolidating Materials — A Status Report,* US Department of Commerce, National Bureau of Standards, May 1980.

Crafts Council, *Science for Conservators, Book 3: Adhesives and Coatings,* Crafts Council Conservation Science Teaching Series, Crafts Council, London, 1983, pp 120–31.

Fry, Malcolm F, 'Exterior Cleaning by Microblasting', *Stone Industries* 18, vol 1, 1983, pp 20–21.

Fry, Malcolm F, 'The Problems of Ornamental Stonework-Graffiti', *Stone Industries,* January/February 1985, pp 26–30.

Garner, Lawrence, *Dry Stone Walls,* Shire Album 114, Shire Publications Ltd, Aylesbury, 1981.

Gauri, K Lal, 'The Preservation of Stone', *Scientific American,* 238, No 6, June 1987, pp 126–36.

Grimmer, Anne E, *A Glossary of Historic Masonry Deterioration Problems and Preservation Treatments,* Preservation Assistance Division, National Park Service, US Department of the Interior, 1984.

Haffey, C and Gray, K, *A Selected Bibliography on Stone Preservation,* Washington DC, 1977 (available from Technical Preservation Services Division, Penston Building, 440 G. Street, N. W, Washington DC 20243).

Haynie, Fred H, 'Deterioration of Marble', *Durability of Building Materials 1,* 1982/1983, pp 241–54.

Heiman, J L, *The Treatment of Salt-Contaminated Masonry With A Sacrificial Render,* Technical Record 471, Experimental Building Station (now National Building Technology Centre), Ryde, NSW, Australia, May 1981.

Heiman, J L, *The Preservation of Sydney Sandstone by Chemical Impregnation* Technical Record 469, Experimental Building Station (now National Building Technology Centre), Ryde, NSW, Australia, March 1981.

Heiman, J L, *Durability Tests on Sandstone Treated with Silanes — Second Series of Tests,* Technical Record 470, Experimental Building Station (now National Building Technology Centre), Ryde, NSW, Australia, April 1981.

Heiman, J L, *The Influence of Temperature and Moisture Movements on Sandstone Masonry,* Technical Record, Experimental Building Station (now National Building Technology Centre), Ryde, NSW, Australia, March 1982.

Honeyborne, D B, *The Building Limestones of France,* HMSO, London, 1982.

Jaynes, S M and Cooke, R V, 'Stone Weathering in Southeast England', *Atmospheric Environment,* vol 21, no 7, 1987, pp 1601–22.

Knoop, D and Jones, G P, *The Mediaeval Mason*, Manchester University Press, 1933; 3rd ed 1967.

Lawson, Judith, *Building Stones of Glasgow*, Geological Society of Glasgow (c/o Geological Department, The University of Glasgow G12 8QQ), 1981.

Leary, Elaine, *The Building Limestones of the British Isles*, HMSO, London, 1983.

Leary, Elaine, *The Building Sandstones of the British Isles*, HMSO, London, 1986.

Lucas, Clive, *Conservation and Restoration of Buildings: Preservation of Masonry Walls*, Australian Council of National Trusts, Canberra, 1982.

Lynch, Michael F and Lynch, William J, *The Maintenance and Repair of Architectural Sandstone*, New York Landmarks Conservancy, New York, 1983.

Martin, D G, *Maintenance and Repair of Stone Buildings*, Church Information Office, London, 1970.

Moncrieff, A and Hempel, K, 'Work on the Degeneration of Sculpture Stone', *Conservation of Stone and Wooden Objects*, IIC, London, 1970, pp 103–14 (Journal of the International Institute for Conservation of Historic and Artistic Works).

Patton, John B, *Glossary of Building Stone and Masonry Terms*, Department of Natural Resources Geological Survey Occasional Paper 6, Authority of the State of Indiana, Bloomington, 1974.

Peterson, S, 'Lime Water Consolidation', *Mortar, Cements and Grouts Used in the Conservation of Historic Buildings*, ICCROM, Rome, 1982, pp 53–61.

Price, C A, *The Decay and Preservation of Natural Building Stone*, BRE Current Paper CP 1/75, Building Research Establishment, Watford, England, 1975.

Price, C A, *Brethane Stone Preservative*, BRE Current Paper CP 1/81, Building Research Establishment, Watford, England, 1981.

Price, C A, 'The Consolidation of Limestone Using a Lime Poultice and Limewater, *Adhesives and Consolidants*, International Institute for Conservation, London, 1984, pp 160–2.

Price, C A and Ross, K D, 'The Cleaning and Treatment of Limestone by the Lime Method; Part II: a Technical Appraisal of Stone Conservation Techniques Employed at Wells Cathedral', *Monumentum*, Winter 1984, pp 301–12.

Purcell, D, *Cambridge Stone*, Faber, 1967.

Rinnie, David, *The Conservation of Ancient Marble*, The J Paul Getty Museum, 1976.

Ritchie, T, 'Roman Stone and Other Defective Artificial Stones', *APT Bulletin* 10, No 1, 1978, pp 20–34.

Ritchie, T, *Silicone Water-Repellents for Masonry*, CBD No. 162, National Research Council, Ottawa, 1974.

Rossi-Manaresi, R, *The Conservation of Stone 1: Proceedings of the International Symposium*, Centro per la Conservazione delle Sculture all'Aperto, Bologna, 1976.

Salzman, L F, *Building in England down to 1540*, Oxford University Press, 1952.

Schaffer, R J, *The Weathering of Natural Building Stones*, Building Research Establishment, HMSO, 1932 (reprinted 1972).

Schaffer, R J, 'Stone in Architecture 1: Stone as a Building Material', reprinted from the *Journal of the Royal Society of Arts*, 103 (4963), 1955, pp 837–67.

Schmedin, G, *Causes and Effects of Surface Deposits, Stains and Decay*, Architectural Conservation Technology Reference System, Vol VI, Parks Canada, Ottawa, 1982.

Simpson, J W and Horrobin, P J, *The Weathering and Performing of Building Materials*, Medical and Technical Publishing Co Ltd, 1970.

Simpson, I M and Broadhurst, F H, *A Building Stones Guide to Central Manchester*, Dept. Extra Mural Studies, University of Manchester, 1975.

Sleater, Gerald A, *Stone Preservatives: Method of Laboratory Testing Preliminary Performance Criteria*, NBS Technical Note 941, National Bureau of Standards, Washington DC, 1977.

Spry, A H, *Principles of Cleaning Masonry Buildings — A Guide to Assist in the Cleaning of Masonry Buildings*, Australian Council of National Trusts, Technical Bulletin 3.1, Canberra, 1982.

Spry, A H, *Chemical Preservation of Sydney Sandstone*, AMDEL Report No 1458, The Australian Mineral Development Laboratories, South Australia, 1983.

Stafford, B F, *Poultice Method for Treating Bituminous Stains on Masonry Products*, Building Research Note 60, National Research Council, Ottawa, 1967.

Stone Federation, *Stone Federation Handbook and Directory*, BEC or Stone Federation, London, 1986. (Includes a glossary of terms in stone masonry.)

Stone Federation, *Becoming A Skilled Stonemason*, Training leaflet.

Stone Industries, published 10 issues per annum by Stone Industries, Ealing Publications Ltd, Weir Bank, Bray, Maidenhead SL6 2ED (Tel: (0628) 23562).

Stone Industries, *Natural Stone Directory*, BEC or Stone Federation, Active Building Stone Quarries, last edition 1987.

Warland, E G, *Modern Practical Masonry*, reprinted by the Stone Federation, London, 1984.

Warnes, A R, *Building Stones*, Benn, 1926.

Winkler, E M, *Stone: Properties, Durability in Man's Environment*, 2nd revised ed, Springer Verlag, Vienna/New York, 1975.

4.5   CLEANING AND POINTING

Many references on masonry also deal with its cleaning and pointing.

Annotated Master Specifications for the Cleaning and Repointing of Historic Masonry, The Ontario Ministry of Citizenship and Culture, Toronto, Canada, 1985.

Ashurst, J, 'The Cleaning and Treatment of Limestone by the "Lime method"' *Monumentum*, Autumn 1984, pp 233–57.

Ashurst, John, 'Cleaning and Surface Repair — Past Mistakes and Future Prospects,' *APT Bulletin*, Vol XV11, No 2, 1985, pp 39–41.

Ashurst, John, *Cleaning Stone and Brick*, Technical Pamplet 4, SPAB, London, 1977.

Ashurst, J, Dimes, F G and Honeyborne, D B, *The Conservation of Building and Decorative Stone*, Butterworth Scientific, London, 1988.

Blades, Keith et al. (eds), *Masonry Conservation and Cleaning Handbook*, APT,

Ottawa, 1984. (Compilation of articles from the US, UK and Canadian sources.)

Building Research Establishment:
Digest 113 *Cleaning External Surfaces of Buildings*, 1983;
Digest 139 *Control of Lichens, Moulds and Similar Growths*, 1982;
BRE, Watford, England.

Clifton, James R (ed), *Cleaning Stone and Masonry*, Symposium, April 1983, ASTM Special Technical Publication 935, ASTM, Philadelphia, PA, 1986.

Grant, C and Bravery, A F, *Laboratory Evaluation of the Water Repellent and Biocide Properties of Organic Compounds*, Building Research Establishment Note 19/86, Princes Risborough Laboratory, Aylesbury, England, February 1986.

Grimmer, Anne E, *Dangers of Abrasive Cleaning to Historic Buildings*, Preservation Brief No 6, Technical Preservation Services Division, Heritage Conservation and Recreation Service, US Department of the Interior, US Government Printing Office, 1979.

Heller, Harold L, 'The Chemistry of Masonry Cleaning', *APT Bulletin*, Vol 2, No IX, 1977, pp 2–9.

Kessler, D W, *A Study of Problems Relating to the Maintenance of Interior Marble*, Technological Papers of the US Bureau of Standards No 350, US Government Printing Office, Washington, 1927.

Mack, Robert C, *The Cleaning and Waterproof Coating of Masonry Buildings*, Preservation Briefs No 1, Technical Preservation Services Division, Heritage Conservation and Recreation Service, US Department of the Interior, US Government Printing Office, 1979.

Mack, Robert C and Askins, James S, *The Repointing of Historic Masonry Building*, SERMAC Industries Inc, Altoona, PA, 1977.

Mack, Robert C. Tiller, de Teel Paterson and Askins, James, *Repointing Mortar Joints in Historic Brick Buildings*, Technical Preservation Services Division, Heritage Conservation and Recreation Service, US Department of the Interior, US Government Printing Office, 1980.

Ritchie, T, *Cleaning of Brickwork*, CBD No 194, National Research Council, Ottawa, 1978.

Spry, A H and West, D G, *The Defence Against Graffiti*, Amdel Report No 1571, AMDEL, Adelaide, South Australia, 1985.

Staehli, A M, 'Appropriate Water Pressures for Masonry Clearing', *APT Bulletin*, Vol XVIII, No 4, 1986.

4.6   STRUCTURAL DEFECTS IN MASONRY

Building Research Establishment:
Digest 75 *Cracking in Buildings* 1966, reprinted 1972;
Digest 200 *Repairing Brickwork*, 1981;
BRE, Watford, England.

Construction Industry Research and Information Association, *Structural Renovation of Traditional Buildings*, Report 111, CIRIA, London, 1986.

Curtis, John Obed, *Moving Historic Buildings*, US Department of the Interior, Heritage Conservation and Recreation Service, Technical Preservation Services Division, US Government Printing Office, Washington DC, 1979.

Heyman, Jacques, *The Masonry Arch*, Ellis Horwood Limited, Chichester, 1982.

Heyman, J, 'The High Endurance of the Masonry Structure', *Design Life of Buildings*, Thomas Telford, London, 1984, pp 59–65.

Macgregor, John E M, *Outward Leaning Walls*, SPAB, Technical Pamphlet No 1, 1971.

Macgregor, John E M, *Strengthening Timber Floors* SPAB, Technical Pamphlet 2, 1973.

Mason, John, 'Vibrating Effects On Buildings', *SPAB News*, Vol 3, No 4, 1982, pp 51–2.

Mason, John, 'Structural Repair – Recording Failure and First Aid Measures', lecture given at SPAB course 'The Repair of Ancient Buildings', London, (no date).

Richardson, Clive, 'Structural Surveys', series of articles:
    'Structural Surveys – Technique and Report Writing', *A J*, 26 June 1985, pp 57–65;
    'Data Sheets: General Problems', *A J*, 3 July 1985, pp 64–71;
    'Data Sheets: The Industrial Revolution', *A J*, 10 July 1985, pp 45–52;
    'Data Sheets: Common Problems 1850–1839' *A J*, 17 July 1985, pp 63–70;
    'Data Sheets: The Post-War Building Boom', *A J*, 24 July 1985, pp 63–70;
    Architects Journal (AJ), The Architectural Press, London.

Steffens, R J, *Structural Vibration and Damage,* Building Research Establishment Report 21-LS-1974, BRE, Watford, England, 1974.

The Assessment of Vibration Intensity and its Application to the Study of Building Vibrations, Building Studies Special Report 19, HMSO, London, 1952.

Torraca, Giorgio, 'Vibration', unpublished paper, ICCROM, Rome, 1977.

Waller, R A, *Building On Springs*, Pergamon Press, 1969.

Williams, G B A, *Chimneys in Old Buildings*, SPAB, Technical Pamphlet No 3, 1976.

4.7   MORTARS, RENDERS, PLASTERS

Adamson, Yvonne, 'The History and Conservation of External Decorative Details In Cements and Plasters', unpublished dissertation, MA, Institute of Advanced Architectural Studies, University of York, 1985.

Ashurst, John, *Mortars, Plasters and Renders in Conservation*, Ecclesiastical Architects' and Surveyors' Association, 1983.

Ashurst, John, 'Mortars for Historic Buildings', *Building Conservation*, Summer 1978.

Bankart, George P, *The Art of the Plasterer*, B T Batsford, London, 1908.

Beard, Geoffrey, *Stucco and Decorative Plasterwork In Europe*, Thames and Hudson Limited, 1983.

British Quarrying and Slag Federation Limited, *Lime In Building*, The Limestone Federation, London, 1968.

British Standards Institution:
    CP 211: 1966, *Internal Plastering*
    CP 231: 1966, *Painting of Buildings*
    BS 1198, 1199, 1200: 1976, *Specifications for Building Sand*
    BS 1191: 1973 Part 1, *Gypsum Building Plasterers*
    BS 890: 1972, *Building Limes*
    BS 5262: 1976, *External Rendered Finishes*
    BS 4721: 1971, *Ready-Mixed Lime: Sand for Mortar*
    BS 5224: 1976, *Specification for Masonry Cement*
    BS 1014: 1975, *Pigment for Portland Cement and Portland Cement Products*
    BS 5390: 1976 (1984), *Code of Practice for Stone Masonry*
    BS 6270: 1982: Part 1: *Code of Practice for Cleaning and Surface Repairs of Buildings*
    BS 4551, *Methods For Testing Mortars, Screeds and Plasters*
Building Research Establishment:
    Digest 160, *Mortars for Masonry*, 1981;
    Digest 196, *External Rendered Finishes*, 1981;
    Digest 213, *Choosing Specifications for Plastering*, 1978;
    BRE, Watford, England.
Building Research Establishment, *Lime and Lime Mortars*, DSIR Special Report No 9 (Cowper), 1927.
Cliver, E Blaine, 'Test for the Analysis of Mortar Samples', *APT Bulletin*, Vol V1, No 1, 1977, pp 68–73.
Crayford, Robert, 'Notes on the Use of Trass in the Wren Period', *ASCHB Transactions*, Volume 5, 1980, pp 11–13.
DOE Advisory Leaflets:
    1 Painting New Plaster and Cement, 1973;
    2 Gypsum Plasters, 1966;
    6 Limes for Building;
    9 Plaster Mixes;
    15 Sands for Plaster, Mortars and Rendering;
    16 Mortars for Brickwork and Blockwork;
    27 Rendering Outside Walls, 1971;
    39 Special Cements;
    57 Newer Types of Paint and Their Uses, 1964;
    Ministry of Public Buildings and Works, Research and Development, HMSO.
ICCROM, *Mortars, Cements and Grouts Used in the Conservation of Historic Buildings*, Symposium, 3–6 November 1981, Rome.
McLaughlin, David, 'Lime-Based Mortars in Bath', *ASCHB Transactions*, Vol 6, 1981, pp 54–5.
Malnic, Nicola, 'Masonry Mortars in Historic Buildings', unpublished dissertation, Master of the Built Environment (Building Conservation), University of NSW, Sydney, 1983.
Millar, William, *Plastering, Plain and Decorative*, Batsford, 1899.
Mora, P, Mora, L and Phillpot, P, *Conservation of Wall Paintings*, Butterworth/ICCROM, 1984.

Newton, R G and Sharpe, J H, 'An Investigation of the Chemical Constituents of Some Renaissance Plasters', *Studies in Conservation*, 32, 1987, pp 163–175.

Pargetting in Hertfordshire, Hertfordshire Conservation File, Hertfordshire County Council, 1983.

Pegg and Stagg, *Plastering – A Craftsman's Encyclopaedia*, Crosby Lockwood Stapels, London, 1976.

Perander, Thorborg and Raman, Tuula, *Ancient and Modern Mortars In The Restoration of Historical Buildings*, Research Notes 450, Technical Research Centre of Finland, Espoo, May 1985.

Phillips, Morgan W, 'Alkali-soluble Acrylic Consolidants for Plaster: A Preliminary Investigation', *Studies in Conservation*, 32, 1987, pp 145–152.

Phillips, Morgan W, 'SPNEA – APT Conference on Mortar', *APT Bulletin*, Vol VI, No 1, 1974, pp 9–39.

Rossage, Mike, 'Coping With Cornices', *Traditional Homes*, Benn Consumer Publications, St Albans, England, April 1986, pp 64–8.

Schofield, Jane, *Basic Limewash*, SPAB Information Sheet 1, London 1986.

Stagg, W.D. and Masters, Ronald, *Decorative Plasterwork – Its Repair and Restoration*, Orion Books, 1983.

Stewart, John and Moore, James, 'Chemical Techniques of Historic Mortar Analysis', *APT Bulletin*, Vol XIV, No 1, 1982, pp 11–16.

Teutonico, Jeanne-Marie, 'Architectural Conservation Course, Laboratory Specifications, Tests and Exercises', unpublished manual, ICCROM, Rome, pp 11–16.

Watson, John and Rastall, R H (ed), *Cements and Artificial Stone – A Descriptive Catalogue of the Specimens in the Sedgwick Museum, Cambridge*, W. Heffer & Sons Ltd, Cambridge, 1922.

4.8 TERRACOTTA AND FAIENCE

Atterbury, Paul and Irvine, Louise, *The Doulton Story*, souvenir booklet produced originally for the exhibition held at the Victoria and Albert Museum, London, 30 May–12 August 1979. (Available from Doulton International Collectors Club Gallery, Leather and Snook, 167 Piccadilly, London W1.

Barry, Charles, 'Some Descriptive Memoranda on the Works Executed in the Terra Cotta at New Alleyn's College, Dulwich, by Mr Blashfield, of Stamford, from the Designs and under the superintendance of Charles Barry, Architect', paper given at the ordinary general meeting of the Royal Institute of British Architects, 22 June 1868.

British Standards Institution:
  BS 6270: Part 1: 1982, *Code of Practice for Cleaning and Surface Repair of Buildings, New Appendix G Cleaning and Surface Repair of Terracotta and Faience*
  BS5385, *Code of Practice for Wall Tiling Part 2: External Ceramic Wall Tiling and Mosaics.*

Collinson, F J, 'Eleanor Coade, 1733–1821' *Museums Journal*, Vol 57, May 1957, pp 37–8.

Cruikshank, Dan and Wyld, Peter, *London: The Art of Georgian Building*, Architectural Press, London, 1975, pp 198–200.

Cupial, Jennie, 'The Coades and Their Stone', *Concrete*, Part 1: October 1980, pp 18–22; Part 2: November 1980, pp 27–30.

Denne, David, 'The Conservation of Decorative Tiling In Building Interiors', unpublished dissertation, MA, Institute of Advanced Architectural Studies, University of York, 1984 (2 vols).

Fidler, John, 'The Conservation of Architectural Terracotta and Faience', *ASCHB Transactions*, Vol 6, 1981, pp 3–16.

Fidler, John, 'The Manufacture of Architectural Terracotta and Faience in the United Kingdom', *APT Bulletin*, Vol XV, No 2, 1983, pp 21–3.

Freestone, I C, Bimson, M and Tite, M S, 'The Constitution of Coade Stone', *Ceramics and Civilisation*, Vol 1, Ancient Techniques to Modern Science, W D Kingery (ed), pp 293–304.

'Glazed Expressions', *Journal of Tiles & Architectural Ceramics Society* (c/o Ironbridge Gorge Museum Trust, Ironbridge, Telford, Shropshire TF8 7AW, UK)

Gunnis, Rupert, *Dictionary of British Sculptors, 1660–1851*, (look under 'Blashfield, 'Coade' and 'Croggan'), Abbey Library, London, 1953.

Hamilton, S B, 'Coade Stone', *Architectural Review*, Vol 116, 1954, pp 295–301.

Howarth, Thomas, 'Tiles, Faience and Mosaic in Modern Building', reprinted from *The Journal of the Royal Institute of British Architects*, August and September 1955.

Kelly, Alison, 'Decorative Stonework in Disguise: Plaques and Medallions in Coade Stone', *Country Life*, Vol 154 (3988), 29 November 1973, pp 1797–8.

Kelly, Alison, 'Coade Stone in Georgian Architecture', *Architectural History*, Journal of the Society of Architectural Historians of Great Britain, Vol 22, pp 71–101.

Lafrance, Marc, 'Coade Stone in Canada', *APT Bulletin*, Vol V, No 3, 1973.

Larney, Judith, *Restoring Ceramics*, Barrie and Jenkins Ltd, London, 1978.

Maltby, Sally, McDonald, Sally and Cunningham, Colin, *Alfred Waterhouse, 1830–1935*, booklet to accompany an exhibition at the RIBA Heinz Gallery, London, 1983.

Neblett, Nathaniel P, 'A Search for Coade Stone in America', *APT Bulletin*, Vol III, No 4, 1977, pp 68–7.

Prudon, T H M, 'Terra Cotta as a Building Material: a Bibliography', *Communique* 5, No 3, APT 1976, Supplement.

Prudon, Theodore H M, 'Architecture Terra Cotta: Analyzing the Deterioration Problems & Restoration Approaches', *Technology & Conservation*, Vol 3, No 3, Autumn 1978, pp 30–8.

Ruch, John E, 'Regency Coade: A Study of the Coade Record Books, 1813–1821', *Architectural History*, Journal of the Society of Architectural Historians of Great Britain, Volume II, 1968, pp 34–56.

Shaws Glazed Brick Co. Ltd, 'Constructional Terra-Cotta and Faience of Today', reprinted from *The Brick Builder*, (England) September 1934.

Snell, Peter, 'The Conservation of Architectural Terracotta', unpublished thesis, Architectural Association, London, Postgraduate Diploma in Building Conservation, 1983.

Stockbridge, Jerry G, 'Analysis of In-Service Architectural Terra Cotta – Support for Technical Investigations', *APT Bulletin*, Vol XVIII, No 4, 1986, pp 41–5.

Stratton, Michael, 'The Terracotta Revival', *Victorian Society Annual*, 1982–3, pp 9–31.

Stratton, Michael, 'The Manufacture and Utilisation of Architectural Terracotta and Faience', PhD Thesis, University of Aston, Birmingham, 1983.

Stratton, Michael, 'The Terracotta Industry: its Distribution, Manufacturing Processes and Products', *Industrial Archaeology Review*, Vol VIII, No 2, Spring 1986, pp 194–214

Summerson, Sir John, *Georgian London*, Penguin, 1945, 1969.

Tiller, de Teel Paterson, *The Preservation of Historic Glazed Architectural Terracotta*, Preservation Brief No 7, Technical Preservation Services Division, Heritage Conservation and Recreation Service, US Department of the Interior, US Government Printing Office, 1979.

Tindall, Susan M and Hamrick, James, *American Architectural Terra-cotta: A Bibliography: (A531)* Vance Bibliographies, Monticello, Illinois, 1981.

Van Lemmen, Hans, *Victoria Tiles,* Shire Album 67, Shire Publications Ltd, Aylesbury, England, 1981.

Ward, Ian, 'The Development, Manufacture, Use and Maintenance of Architectural Terracotta In English Building', Postgraduate Diploma in Architectural Building Conservation, Leicester Polytechnic School of Architecture, 1982–3.

Warren, Charles, 'Notes on Standard Form of Specifications for Architectural Terra-Cotta', *The Brickbuilder* (US), Vol XVI, 1905, pp 8–17.

'W J Neatby's Work and a New Process', *Proceedings of the Society of Designers*, Vol XXV, 1899, pp 88–97.

4.9   DAMP IN MASONRY

Ashurst, Nicola and Williams, Gilbert, 'An Introduction to the Treatment of Rising Damp in Old Buildings', *SPAB Technical Information Sheet*, Winter 1987, pp 9–12.

Beckwijt, W O and B H, 'Measuring Methods for Determining Moisture Content and Moisture Distribution in Monuments', *Studies in Conservation*, May 1970, pp.81–93.

Bowley, M J, *Desalination of Stone: A Case Study*, BRE Current Paper Series, CP 46/75, BRE, Watford, England, 1975.

British Chemical Dampcourse Association (BCDA):
     TIC 1 *The Use of Moisture Meters to Establish the Presence of Rising Damp*;
     TIC 2 *Plastering in Association with Damp-proof Coursing*;
     TIC 3 *Condensation*;
     TIC 4 *Methods of Analysis for Damp-proof Course Fluids*;
     TIC 5 *Chemical Damp-proof Course Insertion – The Attendant Problems*;
     TIC 6 *Safety in Damp Proofing*;
     TIC 7 *Chemical Damp-proof Courses in Walls – Detection Techniques and Their Limitations*;

Code of Practice for Installation of Chemical Damp-proof Courses; BCDA, Berkshire, England.

British Standards Institution, BS 743: 1970, *Specification for Materials for Damp-Proof Courses.*

Building Research Establishment, Technical Information Leaflets,
 TIL 29 *Diagnosis of Rising Damp*, 1977;
 TIL 35 *Electro Osmotic Damp-proofing*, 1975;
 TIL 36 *Chemical Damp-proof Courses for Walls*, 1972;
 TIL 47 *Rising Damp — Advice to Owners Considering Remedial Work*, 1976;
 BRE, Watford, England.

Building Research Establishment:
 Digest 27 *Rising Damp in Walls*, 1962;
 Digest 77 *Damp Proof Courses*;
 Digest 91 *Condensation*, 1969;
 Digest 110 *Condensation*, 1972;
 Digest 177 *Decay and Conservation of Stone Masonry*, 1975;
 Digest 245 *Rising Damp in Walls: Diagnosis and Treatment*, 1981;
 BRE, Watford, England.

Building Research Establishment, *Building Defects and Maintenance*, The Construction Press, London, 1977.

Castle, R W, *Damp Walls*, Technical Press, London, 1964.

Crocker, C R, *Moisture and Thermal Condensation in Basement Walls*, Canadian Building Digest No 161, National Research Council, Ottawa, 1974.

'Damp Proof Course', *Architects Journal*, 21 July 1976, pp 139–41.

De Guichen, Gael, *Climate In Museums*, ICCROM, Rome, 1984.

Duell, John and Larson, Fred, *Damp Proof Course Detailing*, The Architectural Press Ltd, London, 1977 (includes lists of Agrément Board certificates for most damp proof course systems).

Gibbons, G S and Gordon, J L, 'Moisture Migration In Masonry Walls', *Rising Damp: Techniques and Treatments in Building Renovation*, Conference, Graduate School of the Built Environment, University of New South Wales, Sydney, 1978, pp 2–16.

Gratwick, R T, *Dampness in Buildings, I*, Crosby Lockwood and Son Ltd, London, 1966.

Heiman, J L, 'An Evaluation of Methods of Treating Rising Damp', *Rising Damp: Techniques and Treatment in Building Renovations*, Conference, Graduate School of the Built Environment, University of New South Wales, Sydney, 1978, pp 17–40.

Heiman, J L, *Rising Damp and its Treatment*, Technical Information Sheet, Heritage Council of New South Wales, NSW Government Printer, Sydney, 1982.

Heiman, J L, Waters, E H and McTaggart, R C, 'The Treatment of Rising Damp', *Architectural Science Review*, December 1973, pp 170–7.

Hughes, Philip, *The Need for Old Buildings to 'Breathe'*, SPAB Information Sheet 4, Society for the Protection of Ancient Buildings, London, 1986.

Massari, Giovanni, *Humidity In Monuments*, ICCROM and Faculty of Architecture, University of Rome, 1977.

Mora, Paolo, *Causes of Deterioration of Mural Paintings*, ICCROM, Rome, 1974.

Morton, W Brown, 'Moisture in Historic Monuments', *APT Bulletin*, Vol VIII, No 2, 1976, pp 2–19.

Newman, A J, 'The Independent Core Method', *Building Science*, Vol 9, 1974, pp 309–13.

Newman, A J, *Improvements of the Drilling Method for the Determination of Moisture Content in Building Materials – BRE CP 22–75*, Building Research Establishment, Watford, England, 1975.

Oliver, Alan C, *Woodworm, Dry Rot and Rising Damp*, Sovereign Chemical Industries Ltd, UK, 1984.

Oxley, Thomas A and Gobert, Ernest G, *Dampness in Buildings*, Butterworths, London, 1983.

Protection of Buildings Against Water From the Ground, BRE Current paper 102/73, Building Research Establishment, Watford, England, 1973.

Richardson, Barry A, *Remedial Treatment of Buildings*, The Construction Press, Lancaster, UK 1981.

Sheen, J W, *Experiments on the Rain Penetration of Brickwork*, Building Research Establishment, Current Paper 33/71, BRE, Watford, England, 1971.

Smith, Baird M, *Moisture Problems in Historic Masonry Walls – Diagnosis and Treatment*, US Department of the Interior, National Park Service, Preservation Assistance Division, US Government Printing Office, Washington DC, 1985.

Thomas, Andrew R, *Treatment of Damp In Old Buildings*, SPAB Technical Pamplet 8, 1981, revised, 1986

Torraca, Giorgio, *Porous Building Materials – Materials Science for Architectural Conservation*, ICCROM, Rome, 1982.

William, G P, *Drainage around Buildings*, Canadian Building Digest No 1562, National Research Council, Ottawa, 1973.

4.10   TIMBER

Association for Preservation Technology, *The Wood Epoxy Reinforcement System Manual (WER)*, APT, Canada, 1979.

Baker, J M, *Biodeterioration Research At PRL*, BRE Information Sheet 1977/21, Princes Risborough Laboratory, Building Research Establishment, Watford, England, 1977.

Berry, R W, *Environmentally Acceptable Contact Insecticides for Use in Wood Preservation*, BRE Princes Risborough Laboratories, Information Sheet 1976/4, Building Research Establishment, Watford, 1976.

Bravery, A F, 'Biodeterioration of Solid and Constructional Timbers', *Chemistry and Industry*, 20 August 1977, pp 675–8.

Bravery, A F, *Mould and Its' Control*, BRE Information Paper 1985/11, Princes Risborough Laboratory, Building Research Establishment, Watford, 1985.

British Standards Institution:
    BS 144: 1973, *Coal Tar Creosote for the Preservation of Timber*
    BS 1282: 1975, *Guide to the Choice, Use and Application of Wood Preservatives*
    BS 3452: 1962, *Copper/Chrome Water-borne Wood Preservatives and their Application*
    BS 3453: 1962, *Fluoride/Arsenate/Chromate/Dinitrophenol Water-borne Wood Preservatives and their Application*
    BS 4072: 1974, *Wood Preservation by means of Water-borne Copper/Chrome/Arsenic Compositions*
    BS 5056: 1974, *Copper Napthenate Wood Preservatives*
    BS 5268: Pt 5, 1977, *Preservative Treatments for Constructional Timber.*
British Wood Preserving Association,
Technical Leaflets:
    1 Fungal Decay in Buildings — Dry Rot and Wet Rot;
    3 Preservative Treatment of Timber;
    4 Methods of Applying Preservatives;
    5 The Preservation of Window Joinery;
    6 Preservative Treatment Against Wood Borers;
    8 The Use of Creosote Oil for Wood Preservatives;
    9 Organic Solvent Wood Preservatives;
    10 The Treatment of Timber with Waterborne Preservatives;
    11 The Treatment of Solid Timber and Panel Products with Flame Retardants;
    13 Water Repellent Wood Preservatives;
    14 Preserving Wood in Home and Garden — A General Guide;
    BWPA, London.
British Wood Preserving Association, Proceedings to Annual Conventions, BWPA, London.
Brunskill, R W, *Illustrated Handbook of Vernacular Architecture,* Faber and Faber
Brunskill, R W, *Timber Building In Britain,* Victor Gollancz, 1985.
Building Research Establishment:
    Digest 296 *Timbers: Their Natural Durability and Resistance to Preservative Treatment,*
    Digest 299 *Dry Rot: Its Recognition and Control;*
    Digest 307 *Identifying Damage by Wood-boring Insects;*
    BRE, Watford, England
Building Research Establishment (Princes Risborough Laboratory), Technical Notes:
    7 Insecticidal Smokes for Control of Wood-boring Insects, 1966, reprinted 1980.
    39 The House Longhorn Beetle, 1969, reprinted 1980.
    44 Decay in Buildings — Recognition, Prevention and Cure, 1969, revised 1977.
    45 The Death-watch Beetle, 1970, reprinted 1979.
    47 The Common Furniture Beetle, 1970, reprinted 1980.
    55 Damage By Ambrosia (Pinhole Borer) Beetles, 1972, reprinted 1979.
    58 'Woodworm' Control In Domestic Roofs By Dichlorvos Vapour, 1972.
    60 Lyctus Powder — Post Beetles, 1975, reprinted 1979.
    Building Research Establishment, Watford, England.

Building Research Establishment (Princes Risborough Laboratory), *Handbook of Hardwoods*, 2nd ed. HMSO, 1972.

Byrne, R O, Lemire, J, Oberlander, J, Sussman G, and Weaver, M (eds), *Conservation of Wooden Monuments*, Proceedings of the ICOMOS Wood Committee IV, International Symposium, Canada, June 1982, ICOMOS Canada and the Heritage Canada Foundation, Ottawa, 1983.

Carthy, Deborah and McWilliam, Colin (eds), *Decorative Wood*, Proceedings of the Symposium held at the Burrell Collection, Scottish Society for Conservation and Restoration, Glasgow, 31 March 1984.

Cartwright, K St G and Findlay, W P K, *Decay of Timber and its Preservation*, BRE Princes Risborough Laboratory, Monograph, HMSO, 2nd ed.

Charles, F W B, 'The Timber-Frame Tradition and Its Preservation', *ASCHB Transactions*, Vol 3, 1978, pp 5–28.

Coleman, G R, *Control of Death-Watch Beetle in Historic Buildings*, BRE Princes Risborough Laboratories, Information Sheet 1975/8, Building Research Establishment, Watford, England, 1975.

Harris, Richard, *Discovering Timber-Framed Buildings*, Shire Album 242, Shire Publications, Aylesbury, England, 1981.

Harrison, Hugh, 'A Hidden Gem – Techniques Involved in the Conservation of a Medieval Painted Roof', *Church Building*, Summer 1986, pp 23–4.

Hewitt, C A, *English Historic Carpentry*, Fillimore, (no date).

Hide, W T, *Advice on Timber Problems – the Contribution of PRL*, Building Research Establishment, Watford, England.

Lea, R Geraldine, *House Longhorn Beetle Survey*, BRE Princes Risborough Laboratory, Information Paper 1982/12, Building Research Establishment, Watford, England, 1982.

Myers, John H, *The Repair of Historic Wooden Windows*, Preservation Brief No 9, Technical Preservation Services Division, Heritage Conservation and Recreation Service, US Department of the Interior, US Government Printing Office, 1981.

Oliver, Alan C, *Woodworm, Dry Rot and Rising Damp*, Sovereign Chemical Industries Limited, UK, 1984.

Phillips, Morgan W and Selwyn, Dr Judith E, *Epoxies for Wood Repairs in Historic Buildings*, Heritage Conservation and Recreation Service, Technical Preservation Services Division, US Department of the Interior, US Government Printing Office, Washington DC, 1978.

Read, S J, *Controlling Death Watch Beetle*, BRE Princes Risborough Laboratory, Information paper 1986/19, Building Research Establishment, Watford, England, 1986.

Richardson, Barry A, *Wood Preservation*, The Construction Press, Lancaster, 1978.

Ridout, Dr B V, 'The Control of Dry Rot', *Church Building*, Summer 1986, pp 27–8.

Savory, J G, *Dry Rot: Causes and Remedies*, Timberlab paper No 44, Princes Risborough Laboratory, Building Research Establishment, HMSO, London.

Stumes, Paul, 'Testing the Efficiency of Wood Epoxy Reinforcement Systems', *APT Bulletin*, Vol 7, No 3, 1975, pp 2–35.

Timber Research and Development Association, *TRADA Publications*, Brochure, 1985 – Wood Information Sheets – Library Bibliographies.

Timber Pests and Their Control TBL/215, TRADA/BWPA, Revised 1984.

Tokyo National Research Institute of Cultural Properties, *International Symposium on the Conservation and Restoration of Cultural Property – The Conservation of Wooden Cultural Property*, 1–6 November, 1982, Tokyo and Saitama, Japan.

White, M G, *The Inspection and Treatment of Houses for Damage By Wood-boring Insects*, Timberlab Papers, Princes Risborough Laboratory, No 33, 1970, revised 1982, Building Research Establishment, Watford, England.

Wilcox, R P, *Timber and Iron Reinforcement In Early Buildings*, Occasional Paper (new Series) II, The Society of Antiquaries of London, London, 1981 (Burlington House, Piccadilly, London W1V 0HS).

4.11   ROOFING

General

Fidler, John, 'Roof Recognition', *Traditional Homes*, March 1985, pp 67–74.

Institute of Plumbing, *Sheet Roofing Data Book and Design Guide*, Technical Committee of Oxford District Council and the Institute of Plumbing, Oxford, 1978.

Lucas, Clive, *Conservation and Restoration of Buildings: Conservation of Roofs*, Australian Council of National Trusts, Canberra, 1979.

'Sheet Roofing Materials, Products in Practice', *AJ* Supplement, 28 March 1984, pp 22–39.

Sweester, Sarah M, *Roofing For Historic Buildings*, Preservation Brief No 4, Technical Preservation Services Division, Heritage Conservation and Recreation Service, US Department of the Interior, US Government Printing Office, 1979.

Thatch

Brockett, Peter and Wright, Adela, *The Care and Repair of Thatched Roofs*, Society for the Protection of Ancient Buildings and Council for Small Industries in Rural Areas, SPAB Technical Pamphlet 10, SPAB, London, May 1986.

Council for Small Industries in Rural Areas, *The Thatcher's Craft*, CoSIRA, Salisbury, 1977.

Fearn, Jacqueline, *Thatch and Thatching*, Shire Album 16, Shire Publications, Aylesbury, 1981.

Thatch in Hertfordshire, Hertfordshire Conservation File, Hertfordshire County Council, 1983.

Thatching Advisory Service Ltd, *Specifications on Thatch – A Guide for the Architect and Builder Dealing with Thatch; Insuring Thatch – A Guide for All Householders of Thatch Properties*, Wokingham, England.

West, Robert, *Thatch: A Manual for Owners, Surveyors, Architects and Builders*, David and Charles, 1987.

Slate and stone

Ballard, Candice, *Working with Slate Roofs*, The Heritage Canada Foundation, 1984.

'A Cotswold Stone Roof,' *AJ*, 16 January 1985, pp 57–63.

British Standards Institution:

    BS 680: 1971: Part 2, *Roofing Slates (Metric Units)*

    BS 5534: 1978: Part 1, *Code of Practice for Slating and Tiling (Design)*

Mineral Resources Consultative Committee, *Slate*, Mineral Dossier No 12, Institute of Geological Sciences, Mineral Resources Division, HMSO, 1975.

Tiles and Slates In Hertfordshire, Hertfordshire Conservation File, Hertfordshire County Council, 1984.

Traditional Stone Roofing, Design and Conservation Section, County Planning Department, Derbyshire County Council.

Watkins, C M and Brady, F L, *The Durability of Slates for Roofing*, Building Research Bulletin No 12, Department of Scientific and Industrial Research, HMSO, 1932.

Shingles

Stewart, Ian, 'Recovering Somptings's Spire', *SPAB News*, Vol 5, No 4, October 1984, pp 58–60.

4.12   METALS

Metals — properties and corrosion

Berry, P, *Corrosion Principles for Engineering Technicians*, Department of Industry, London, 1982.

British Standards Institution:

    PD 6484: 1979, *Commentary on Corrosion at Bimetallic Contacts and its Alleviation*

    Draft BS 6270: Part 3, *Code of Practice for Cleaning and Surface Repair of Buildings, Part 3 — Metals*

British Steel Corporation, The Prevention of Corrosion on Structural Steels, BSC Sections, Cleveland, UK (no date).

Building Research Establishment:

    Digest 301, *Corrosion of Metal by Wood*, September 1985;

    Digest 71, *Painting in Buildings: 2, Non-ferrous Metals and Coatings*, 1969;

    Digest 110, *Corrosion of Non-ferrous Metals: I*, 1958;

    Digest 111, *Corrosion of Non-ferrous Metals: II*, 1958;

    BRE, Watford, England.

Button, Harry, 'The Restoration of Eros', *London Environmental Bulletin*, Vol 2, No 3, Winter 1984/5, p 2.

CoSIRA, *Metals for Engineering Craftsmen*, Council for Small Industries in Rural Areas, Salisbury, 1979.

Department of Industry and Institute of Corrosion Science and Technology;

Guides to Practice in Corrosion Control:

    9 *Cathodic Protection*, 1981;

12 *Paint for the Protection of Structural Steelwork*, 1981;

13 *Surface Preparation for Painting* 1982;

14 *Bimetallic Corrosion* 1978;

DI and COI, London.

Evans, Ulick R, *An Introduction to Metallic Corrosion*, Edward Arnold Ltd, London, 1st ed 1948, 2nd ed 1963.

Gayle, Margot, Look, David W, and Waite, John G, *Metals in America's Historic Buildings — Uses and Preservation Treatments*, Technical Preservation and Recreation Service, US Department of the Interior, US Government Printing Office, Washington DC, 1980.

Geerlings, Gerald K, *Metal Crafts in Architecture*, Bonanza Books, New York, 1957.

Hughes, Richard and Rowe, Michael, *The Colouring, Bronzing and Patination of Metals*, Crafts Council, London, 1982.

Nielsen, N A, 'Corrosion Product Characterization', *Corrosion and Metal Artefacts — A Dialogue Between Conservator, Archaeologists and Corrosion Scientists*, National Bureau of Standards, Washington DC, 1977, pp 17–37.

Organ, R M, 'The Corrosion of Tin, Copper, Iron, Steel and Lead' *Preservation and Conservation: Principles and Practices*, Sharon Timms (ed), Conference 10–16 September 1972, Preservation Press, Washington DC, 1976, pp 243–56.

Percy, John, *Metallurgy — The Art of Extracting Metals From Their Ores, and Adapting Them to Various Purposes of Manufacture*, First printed, John Murray, Albemarle Street, London, 1864, Facsimile reprint by De Archaeologische Pres, Nederland:

Volume I — Part 1 Fuel, Fire Clays; Part 2 Copper, Zinc, Brass;

Volume II — Part 1 Properties of Iron, Iron Ores, Direct Reduction Processes; Part 2 Iron: Indirect Reduction Processes;

Volume III — Part 1 Lead Desilverisation; Part 2 Lead;

Volume IV — Part 1 Silver; Part 2 Gold.

Rivington's Notes on Building Construction, 'Part 3: Materials', Longman Green & Co., London, 1982.

Scottish Society for Conservation & Restoration, *The Conservation and Restoration of Metals*, Proceedings of the Symposium, Edinburgh, 30–31 March 1979.

Shreir, L L, *Corrosion: Metal/Environment Reactions*, 2nd ed, 2 vols, Newnes — Butterworth, London, 1976.

Street, Arthur and Alexander, William, *Metals in the Service of Man*, Pelican, Middlesex, first published 1944, reprinted 1954.

Twopenny, William, *English Metalwork*, Architectural Constable & Co, London, 1904.

Weil, Phoebe Dent, 'Conservation of Metal Statuary and Architectural Decoration in Open-Air Exposure: An Overview of Current Status with Suggestions regarding Needs and Future Direction', ICCROM Conference, *Conservation of Metal Statuary and Architectural Decoration in Open-Air Exposure*, Paris, 6–8 October, 1986.

Wranglen, Gosta, *Corrosion and Protection of Metals*, Chapman and Hall, London, 1985.

Cast iron and wrought iron

Allen, Nicholas K, 'The Care and Repair of Cast and Wrought Iron', Unpublished dissertion, Diploma in Conservation Studies, University of York, 1979.

Aston, James and Story, Edward B, *Wrought Iron — Its Manufacture, Characteristics and Applications*, Am Byers Co., Pittsburgh, Pennysylvania, 1939.

Barraclough, K C, *The Origin of the British Steel Industry*, Sheffield City Museums, Information Sheet 7.

Barraclough, K C, *Crucible Steel Manufacture*, Sheffield City Museums, Information Sheet 8.

Barraclough, K C, *Bessemer and Sheffield Steelmaking*, Sheffield City Museums, Information Sheet 18.

Bidwell, T G, 'The Restoration and Protection of Structural and Decorative Cast Iron at Covent Garden Market', *ASCHB Transactions*, Vol 5, 1980, pp 24—30.

Blackshaw, S M, 'An Appraisal of Cleaning Methods for Use on Corroded Iron Antiquities', *Conservation of Iron*, National Maritime Museum, Maritime Monographs and Reports No 53, 1982, pp 16—22.

British Standards Institution:
   BS 1452: 1961, *Specification for Grey Iron Castings*
   BS 2569: Part 1: 1964, *Protection of Iron and Steel by Aluminium and Zinc against Atmospheric Corrosion*
   BS 54732: 1967, *Surface Finish of Blast-Cleaned Steel for Painting*
   BS 5493: 1977, *Code of Practice for Protective Coating of Iron and Steel Structures against Corrosion*
   CP 2008: 1966, *Protection of Iron and Steel Structures from Corrosion*
   CP 3012: 1972, *Cleaning and Preparation of Metal Surfaces*

British Steel Corporation, *The Prevention of Corrosion on Structural Steels*, Technical Note, BSC Sections, Cleveland, UK (no date).

Brough, Joseph, *Wrought Iron — The End of an Era at Atlas Forge Bolton*, Bolton Metropolitan Borough Arts Department, late 1970s.

Building Research Establishment, Digest 70 *Painting: Iron and Steel*, BRE, Watford, England, 1973.

Canning, W, plc *The Canning Handbook of Surface Finishing Technology*, W Canning plc, Birmingham, in association with E & F N Spon Ltd, London, 1985.

'Cast Iron's Potential', *A J*, London, 4 July 1984.

Clarke, R W and Blackshaw, S M (eds), *Conservation of Iron*, National Maritime Museum, Maritime Monographs and Reports No 53, 1982.

CoSIRA, *Wrought Ironwork — A renewal of Instruction for Craftsmen*, Publication No 55, Council for Small Industries in Rural Areas, Salisbury, England, 8th impression, January 1981.

CoSIRA, *The Blacksmith's Craft — An Introduction to Smithing for Apprentices and Craftsmen*, Council for Small Industries in Rural Areas, Salisbury, England, 10th impression, August 1983.

Council of Ironfoundry Associates, *A Guide to the Engineering Properties of Castings*,

The Joint Iron Council, 14 Pall Mall, London SW1, TC of IA, London, 1964.

Crafts Council, *Science for Conservators, Book 2: Cleaning*, Crafts Council Conservation Science Teaching Series, Crafts Council, London, 1983.

Department of Industry and Institute of Corrosion Science and Technology, Guides to Practice in Corrosion Control:

  9 Cathodic Protection, 1981;

  12 Paint for the Protection of Structural Steelwork, 1981;

  13 Surface Preparation for Painting, 1982;

  14 Bimetallic Corrosion

  Department of Industry and Institute of Corrosion Science and Technology, London.

Edwards, Ifor, *Davies Brothers, Gatesmiths — 18th Century Wrought Ironwork In Wales*, Welsh Arts Council/Crafts Advisory Committee, Cardiff, 1977.

Ekey, D C and Winter, W P, *Introduction to Foundry Technology*, McGraw Hill Book Company, New York, 1958.

Evans, V R, *The Rusting of Iron: Cause and Control*, Edward Arnold, London, 1972.

Fairburn, W, *The Application of Cast and Wrought Iron to Building Purposes*, London, Longmans, 1864.

Gale, W K V, many references on the history of the British Iron Industry, e.g. *Ironworking*, Shire Album 64, Shire Publications Ltd, Aylesbury, England, 1981.

Geddes, Jane, 'Medieval Decorative Ironwork 1100–1350', unpublished dissertation, Courtauld Institute, London University, 1978.

Geerlings, Gerald K, *Wrought Iron in Architecture — An Illustrated Survey*, Dover Publications, New York, 1929.

Gilbert, G N J, *Engineering Data on Grey Cast Irons*, British Cast Iron Research Association, 1968.

Greater London Council, Materials Bulletin No 91, *Cast Iron Columns and Beams*, out of print.

Harding, Richard, 'Engineers Cast New Roles for Iron', *New Scientist*, 6 March 1986, pp 40–3.

Hawkes, Pamela W, 'Paints for Architectural Cast Iron', *APT Journal* XI, 1979, pp 17–35.

Howard, E D, *Modern Foundry Practice*, Oldhams Press Limited, Long Acre, London.

Hume, Ian, 'The Iron Bridge, Shropshire: Repainting and Repairs 1980', *ASCHB Transactions*, Vol 5, 1980, pp 20–23.

The Institute of Industrial Archaeology, *International Seminar On Wrought Iron*, Proceedings of the symposium held at Ironbridge, 14–17 July 1986, the University of Birmingham and the Ironbridge Gorge Museum Trust.

Ironbridge Gorge Museum Trust, Museum Guides:

  2.01 *The Coalbrookdale Museum of Iron*, 1979;

  3.01 *The Iron Bridge*, 1979;

  3.02 *Bedlam Furnace*, 1984;

  4.01 *Blists Hill Open Air Museum*, 1985;

20.02 *Coalbrookdale Ironworks*, 1975;

20.03 *Coalbrookdale in 1801*, 1979;

20.04 *Iron and Steel*, (W K V Gale) 1979;

Information Sheet 4, *The Coalbrookdale Company and the SS Great Britain*; Ironbridge Gorge Museum Trust, Ironbridge, Telford, Shropshire, TF8 7AW.

Ironworks and Iron Monuments — Forges et Monuments en Fer, Publication of the proceedings of a conference held in Ironbridge, Shropshire, ICCROM, 22–25 October 1984.

Jones, Chris, 'Two Glass Houses — Syon and Bicton', *AJ*, 29 April 1987, pp 57–62.

Jones, S B and Gregory, E N, 'The Welding of Cast Irons', reprinted from *Welding Institute Research Bulletin* R103/12/73, 1973.

King, Michael, 'The Restoration and Repair of Cast Iron and Glass Verandahs in Lord Street, Southport, Merseyside', *ASCHB Transactions*, Vol 6, 1981, pp 17–22.

Lister, Raymond, *Decorative Cast Ironwork in Great Britain*, G Bell and Sons Limited, London, 1960.

McLaughlin, David, 'The Repair of a Cast Iron Bridge Over the Kennet and Avon Canal, Sydney Gardens, Bath', *ASCHB Transactions*, Vol 5, 1920, pp 30–3.

Paint Research Association, *Quality Control Procedures when Blast Cleaning Steel*, PRA, Teddington, November 1980.

'Pride of Iron — Restoring the Palm House' (Kew Gardens), *Construction 58*, pp 23–8, Journal of the Property Services Agency, England, 1987.

Rivington's Series of Notes on Building Construction, *Notes on Building Construction, Part 3, Materials*, Chapter IV: Metals/Cast Iron, Wrought Iron and Steel, Longman, Green & Co., London, 1892.

Shelton, Roderick, 'Architectural Cast Iron', unpublished thesis, Architectural Association, London, Postgraduate Diploma in Building Conservation, May 1982.

Squires, G A, *Institute's Advice Saves Station Roof*, (Welding Wrought Iron), available from The Welding Institute, Cambridge.

Starkie Gardiner, J, *Ironwork: Part I: From the Earliest Time to the End of the Medieval Period; Part II: Continental Ironwork of the Renaissance and Later Periods; Part III: The Artistic Working of Iron in Great Britain From the Earliest Times*; London, first published 1892, reprinted by the V & A Museum, 1978.

Trinder, Barrie, several books and articles on iron and its history in Britain, e.g. 'The Use of Iron as a Building Material', *Timber, Iron, Clay — Five Essays on Their Use in Building*, D Linstrum, A Clifton-Taylor, M Opie, B Trinder, R Brunskill, West Midlands Arts Publications, pp 38–49.

Walker, R, 'The Role of Corrosion Inhibitors in the Conservation of Iron' *Conservation of Iron*, National Maritime Museum, Monograph No 53, 1982, pp 58–65.

Watkinson, F and Boniszewski, T, 'The Selection of Weld Metal for Wrought Iron', British Welding Research Association, *BWRA Bulletin*, Vol 6, No 9, September 1965.

Williams, Sarah, 'Traditional Wrought Iron Details From the 17th, 18th and 19th

Centuries', consultant study prepared for the Research and Technical Advisory Service, English Heritage, 1987 (unpublished).

Yeomans, D, 'New Life for Cast Iron', *AJ*, 7 May 1986.

Lead

Andre, J Lewis, 'English Ornamental Leadwork', *Archaeological Journal*, Vol 45, 1888, pp 109–19.

British Standards Institution:

    BS 1178, 1982, *Milled Lead Sheet and Strip for Building Purposes*

    CP143 Part 2, *Draft Specification for the Design and Construction of Fully Supported Lead Sheet Roof and Wall Coverings* (Draft out for public comment)

    BS 6229: 1982, *Flat Roofs with Continuously Supported Coverings.*

Burt, Roger, *The British Lead Mining Industry*, Dyllansow Truran Trevolsta, Trewirgie, Redruth, Cornwall, 1984.

Ecclesiastical Architects and Surveyors Association 'Corrosion of Lead Roofing', in preparation (1988).

Harn, O C, *Lead: The Precious Metal*, The Century Co, London, 1924.

Hoffmann, Wilhelm, *Lead and Lead Alloys; Properties and Technology*, Springer Verlag, Berlin, 1970

Jackson-Stops, Gervase, 'New Deities for Old Parterres – The Painting of Lead Statues', *Country Life*, 29 January 1987, pp 92–4.

Krysko, Vladimir K, *Lead In History and Art*, Dr Riederer Verlag GmbH, Stuttgart, 1979.

Lead Development Association, *Lead Sheet in Building: A Guide to Good Practice*, LDA, London, 1978 (revised 1984).

Lead Development Association, *Lead Sheet Flashings for Slate and Tile Roofing*, LDA, London, March 1981.

Lead Development Association, *Leadwork*, quarterly journal since September 1979, LDA, London.

Lead Development Association, *Lead Contractors Association Newsletter No 9 – Condensation Can Bring Corrosion*, LDA, London, 16 October 1986.

Lethaby, W R, *Leadwork*, Macmillan & Co., London and New York, 1893.

Melville, Rodney, 'Non-traditional Material in Church Roofing', *Conference on New Materials in The Conservation of Churches*, Council for the Care of Churches, November 1981, pp 8–17.

Naylor, Andrew, 'Lead Statues: Their Conservation and Presentation', *Historic Houses*, Vol 7, No 3, October 1983, pp 35–9.

Naylor, Andrew, 'Naylor Conservation – Recent Work of a Specialist Conservation Service', ICCROM Conference *Conservation of Metal, Statuary and Architectural Decoration in Open-Air Exposure*, Paris, 6–8 October, 1986.

Naylor, Janet, 'The Conservation and Restoration of Lead Based Statuary', *Scottish Society for Conservation and Restoration Bulletin*, 1983, pp 9–11.

Rowe, D J, *Lead Manufacturing In Britain*, Croom Helm, London & Canberra, 1983.

'Sheet Roofing Materials – Products in Practice', *AJ* Supplement, 28 March 1984, pp 23–39.

Simpson, Lorne Gordon, 'Conservation of Ornamental Leadwork', MA Conservation Studies, University of York, Institute of Advanced Architectural Studies, August 1986.

Smythe, J A, *Lead*, Longmans, Green and Co., London, 1923.

Smythe, J A, *Pitman's Common Commodities and Industries: Lead*, Sir Isaac Pitman and Sons Ltd, Bath, Melbourne and New York, 1970.

Troup, F W, 'Ornamental Lead and Lead-Casting', *RIBA Journal*, 3rd Series, Vol. VIII, No 13, 12 May 1900, pp 327–39.

Troup, F W and Weaver, Lawrence, 'Leadwork', *RIBA Journal*, 3rd Series, Vol. XIII, No 10, 24 March 1906, pp 257–70.

Weaver, Lawrence, 'English Lead Fonts', *Architectural Review*, Vol XIX, No 112, March 1906, pp 99–108.

Weaver, Lawrence, 'Lead Garden Statues' *Architectural Review*, Vol XX, No 117, August 1906, pp 60–75.

Weaver, Lawrence, *English Leadwork: Its Art and History*, B T Batsford, London, 1909.

Willies, Lynn, *Lead and Lead Mining*, Shire Album 85, Shire Publications, Aylesbury, England, 1982.

Zarnecki, George, *English Ornamental Leadwork*, Alec Tiranti, London, 1957.

Copper

Atkinson, R L, *Copper and Copper Mining*, Shire Album 201, Shire Publications, Aylesbury, England, 1987.

British Standards Institution:
CP 143 *Sheet Roof and Wall Coverings, Part 12: 1970 Copper, Metric Units*
BS 2870: 1960 *Rolled Copper Sheet, Strip and Foil*, metric.
British Standard Code of Practice 3, Chapter V, Part 2, 1972.

Cliver, E Baine, 'The Statue of Liberty: A Monument to Metal' ICCROM Conference *Conservation of Metal Statuary and Architectural Decoration In Open-Air Exposure*, Paris, 6–8 October, 1986.

Copper Development Association, *Copper Through the Ages*, CDA Publication No 3, first issued 1934, revised 1956.

Copper Development Association, *Copper in Roofing – Design and Installation*, TN 32 CDA, Potters Bar, December 1985.

Glover, H, 'The Development of Light Gauge Sheet and Strip for Roofing and Cladding', paper given at *Sheet Roofwork and Weathering Symposium*, conference organized by the Institute of Plumbing, Oxford College of Further Education, 1 April 1987.

Hemming, D C, 'The Production of Artificial Patination On Copper', *Corrosion and Metal Artefacts – A Dialogue Between Conservators and Archaeologists and Corrosion Scientists*, National Bureau of Standards Special Publication 479, July 1977, pp 93–102.

Institute of Plumbing, *Sheet Roofing Data Book and Design Guide*, Technical Committee, Oxford District Council and the Institute of Plumbing, Oxford, 1978.

Bronze

Chase, W T and Veloz, Nicholas F, *Some Considerations in Surface Treatment of Outdoor Metal Sculptures* The American Institute for Conservation of Historic and Artistic Works (AIC), preprints of papers presented at the thirteenth annual meeting, Washington DC, 22–26 May 1985.

Fry, M F, 'Exterior Cleaning by Microblasting', *Stone Industries* 18, Vol 1, 1983, p 20.

Gayle, Margot, Look, David W and Waite, John G, *Metals in America's Historic Buildings — Uses and Preservation Treatments*, Heritage Conservation Recreation Service, Technical Preservation Services Division, US Department of the Interior, US Government Printing Office, Washington DC, 1980.

Generic Guidelines for the Preservation of Bronze Statuary, Plaques and Other Bronze Architectural Features in the Southeast Region, Preservation Assistance Service (PAS), US Department of the Interior, National Park Service, 1984.

Hamilton, H, *The English Brass and Copper Industries to 1800*, Frank Cass & Co. Ltd, London 1st ed 1926, 2nd ed 1967.

Hughes, Richard and Rowe, Michael, *The Colouring, Bronzing and Patination of Metals*, Crafts Council, London, 1982.

Jack, J F S, 'The Cleaning and Preservation of Bronze Statues', *The Museums Journal*, Vol 50, No 10, January 1951, pp 231–5.

La Fontaine, Raymond H, 'The Use of a Stabilizing Wax to Protect Brass and Bronze Artifacts', *Journal of the International Institute for Conservation — Canadian Group* Vol 4, No 2, short contribution.

Madsen, H B, 'A Preliminary Note on the Use of BTA for Stabilising Bronze Objects', *Studies in Conservation*, Vol 12, 1967, pp 163–7.

Marabelli, Maurizio, 'Characterization and Conservation Problems of Outdoor Metallic Monuments', Istituto Centrale del Restauro, Rome, paper given at ICCROM Conference, *Conservation of Metal Statuary and Architectural Decoration in Open-Air Exposure*, Paris, 6–8 October 1986.

Merk, L E, 'A Study of Reagents Used in the Stripping of Bronzes', *Studies in Conservation*, No 23, 1978, p 15.

Morris, K and Krueger, J W, 'The Use of Wet Peening in the Conservation of Outdoor Bronze Sculpture', *Studies in Conservation*, No 24, 1979, p 40.

Oddy, W A, 'On the Toxicity of Benzotriazole', *Studies in Conservation*, No 17, 1971, p 135.

Oddy, W A, 'Toxicity of Benzotriazole', *Studies in Conservation*, No 19, 1974, p 188.

Organ, R M, 'Aspects of Bronze Patina and Its Treatment', *Studies in Conservation*, Vol 8, No 1, February 1983, pp 1–9.

Savage, George, *A Concise History of Bronzes*, Thames and Hudson Ltd, London, 1968.

Sease, C, 'Benzotriazole: A Review for Conservators', *Studies in Conservation*, No 23, 1978, p 76.

Sivinski, Valerie, 'Conservation Handbook for Copper and Copper Alloy Architectural Ornamentation', MA in Conservation Studies, University of York, Institute of Advanced Architectural Studies, August 1986. (Contains a very good bibliography.)

Smith, Rika and Beale, Arthur, 'An Evaluation of the Effectiveness of Various Plastic and Wax Coatings in Protecting Outdoor Bronze Sculpture Exposed to Acid Deposition: A Progress Report', ICCROM Conference, *The Conservation of Metal Statuary and Architectural Decoration in Open-Air Exposure*, Paris, 6–8 October 1986.

Stambolov, T, 'Introduction to the Conservation of Ferrous and Non-ferrous Metals', *The Conservation and Restoration of Metals*, Scottish Society for Conservation and Restoration, proceedings of the symposium, Edinburgh, 1979, pp 10–19.

Walker, R, 'The Role of Benzotriazole in the Preservation of Antiquities', *The Conservation and Restoration of Metals*, Scottish Society for Conservation and Restoration, proceedings of the symposium, Edinburgh, 1979, pp 40–9.

Weil, Phoebe Dent, 'The Use of Glass Bead Peening to Clean Large-Scale Outdoor Bronze Sculpture', *American Institute for Conservation Bulletin*, Vol 15, No 1, 1974, p 51.

Weil, Phoebe Dent, 'A Review of the History and Practice of Patination' *Corrosion and Metal Artifacts — a Dialogue Between Conservators and Archaeologists and Corrosion Scientists*, P F Brown et al (eds) National Bureau of Standards, Special Publication 479, US Department of Commerce, July 1977.

Weil, Phoebe Dent, 'The Corrosive Deterioration of Outdoor Bronze Sculpture', *Science and Technology in the Service of Conservation*, Brommelle, N S, and Thomson, G (eds), International Institute for Conservation, London, 1982, p 130.

Weil, Phoebe Dent, *Maintenance Manual for Outdoor Bronze Sculpture*, Washington University Technology Associates (WUTA), St Louis, Missouri, January 1983.

Zinc

British Standards Institution:

CP 143 *Sheet Roof and Wall Coverings, Part 5: Zinc'*

BS 849: *Plain Zinc Sheet Roofing* (superseded by BS 6561: 1985)

BS 6561: 1985, *Plain Zinc Sheet Roofing*

BS 2717, *Glossary of Terms Applicable to Roofing*

Building Research Establishment, Digest 71, *Painting of Buildings: 2, Non-Ferrous Metals and Coatings*, BRE, Watford, England, 1969.

Gayle, Margot, Look, David W and Waite, John G, *Metals in America's Historic Buildings — Uses and Preservation Treatments*, Technical Preservation Services Division, Heritage Conservation and Recreation Service, US Department of the Interior, US Government Printing Office, 1980.

Institute of Plumbing, *Sheet Roofing Data Book and Design Guide*, Technical Committee, Oxford District Council and the Institute of Plumbing, Oxford, 1978.

Ridge, George H, *Paper No. 2*, Symposium held by the Institute of Plumbing, Oxford District Council, Oxford, April 1978.

Ridge, George H, 'History of Zinc Roofing', *Roofing and Building Insulation: Development Through 150 Years — Impetus from Railway Growth*, Issue 3, March 1958, pp 2–3; 'Nineteenth Century Developments — Its Decorative Value, Issue 4, 1958, pp 10–11; 'History of Zinc Roofing — 1920 to the Present Day', Issue 5, 1958, pp 10–11.

Zinc Development Association, *Zinc in Building:* Data Sheet 1 'Roll Cap Roofing'; Data Sheet 2 'Standing Seam Roofing; ZDA, August 1971.

Zinc Development Association, *Metizinc*, ZDA, January 1986. (Modern Zinc Sheet Roofing Practice).

Zinc sculpture

Gayle, Margot, Look, David W and Waite, John G, *Metals in America's Historic Buildings — Uses and Preservation Treatments*, Technical Preservation Services Division, Heritage Conservation and Recreation Service, US Department of the Interior, US Government Printing Office, 1980.

Naylor Conservation, 'Report No 191 — Conservation Report on the Zinc Statues at Osborne House, Isle of Wight', Report for RTAS, English Heritage, unpublished, Survey date: February 1987, Naylor Conservation, Telford, Shropshire.

Nosek, Elzbieta Maria, 'Conservation of Outdoor Zinc Monuments' ICCROM Conference *Conservation of Metal Statuary and Architectural Decoration in Open-Air Exposure*, Paris, 6–8 October, 1986.

Weil, Phoebe Dent, 'Problems in the Conservation of Zinc Sculpture in Outdoor Exposure', ICCROM Conference *Conservation of Metal Statuary and Architectural Decoration in Open-Air Exposure*, Paris 6–8 October, 1986.

Tin

Atkinson, R L, *Tin and Tin Mining*, Shire Publications, Aylesbury, England, 1985.

4.13   PAINT

Bristow, Ian, 'The Redecoration of the Dulwich Picture Gallery, 1980–81', *ASCHB Transactions*, Vol. 6, 1981, pp 33–6.

British Standards Institution, CP 231: 1966, *Painting of Buildings.*

Building Research Establishment:
Digest 197 *Painting Walls: Choice of Paint*, 1982;
Digest 21 *New Types of Paint*;
Digest 71 *Paintings in Buildings: 2, Non-ferrous Metals and Coatings*, 1969;
BRE, Watford, England.

DOE Advisory Leaflets:
11 Painting Metalwork, 1970;
25 Painting Woodwork, 1973;
37 Emulsion Paints, 1973;

57 *Newer Types of Paint and Their Uses*, 1964;
Ministry of Public Buildings and Works, HMSO.

Exterior Painting, Special Issue: *The Old House Journal*, Vol. 4, No 4, April 1981, pp 71–94.

Feller, Robert L, 'The Deterioration of Organic Substances and the Analysis of Paints and Varnishes', *Preservation and Conservation: Principles and Practices*, Sharon Timms (ed), 1976, pp 287–99.

Gehrig, Keith and Ellsmore, Don (ed), *Research Study No 9 – A Guide to Traditional Painting Techniques*, The Heritage Council of New South Wales, Sydney, 1985.

Hamburgh, J R and Morgan, W H (eds), *Hess's Paint Film Defects, Their Causes and Cures*, 3rd ed., Chapman & Hall, London, (no date).

Innes, Jocasta, *Paint Magic*, Windward, Berger Paints, 1981.

Jolly, V G, *An Introduction to Paint*, The Walpamur Co Ltd, W S Cavell Ltd, 2nd ed, 1959.

MacTaggart, Peter and Ann, *Practical Gilding*, Mac & Me Ltd, Welwyn, Herts, England, 1984.

Miller, Kevin H, *Paint Colour Research and Restoration of Historic Paint*, Association for Preservation Technology, Ottawa, 1977.

Moss, Roger, *Century of Colour: Exterior Decoration for American Buildings, 1820–1920*, The American Life Foundation, Watkins Glen, New York, 1981.

Phillips, Morgan W, 'Problems in the Restoration and Preservation of Old House Paints, *Preservation and Conservation: Principles and Practices*, Sharon Timms (ed), 1976, pp 273–85.

Phillips, Morgan W, 'Acrylic Paints for Restoration – Three Test Application' *APT Bulletin*, Vol XV, No 1, 1983, pp 3–11.

Schofield, Jane, *Basic Limewash*, SPAB Information Sheet 1, London, 1986.

Weeks, Kay D and Look, David W, *Exterior Paint Problems on Historic Woodwork*, Preservation Briefs No 10, Technical Preservation Services, Preservation Assistance Division, National Park Service, US Department of the Interior, 1982.

Wright, Adela, *Removing Paint From Old Buildings* SPAB Information Sheet 5, *SPAB News*, Summer 1986.

4.14   EARTH WALLS

Alva, Alejandro and Giacomo, Chiari, 'Protection and Conservation of Excavated Structures of Mudbrick', unpublished paper, ICCROM, 1982.

Alva, Alejandro and Teutonico, Jeanne-Marie, 'Notes on the Manufacture of Adobe Blocks for the Restoration of Earthern Architecture', unpublished paper, Rome, ICCROM, 1982.

British Standards Institution, BS 1377: 1975, *Methods of Test for Soil for Civil Engineering Purposes*.

Brunskill, J, *Illustrated Handbook of Vernacular Architecture*, Faber and Faber, 3rd edition, 1987.

Clifton-Taylor, Alec, *The Pattern of English Building*, Batsford, London, 1962; Faber & Faber 1972.

Department of Housing and Urban Development (US), *Handbook for Building Homes of Earth,* Office of International Affairs, Washington DC.

Experimental Building Station:

Bulletin No 5, *Earth Wall Construction* (1981);

Notes on the Science of Building:

NSB 13, *Earth-Wall Construction* (1971);

NSB 18, *Pise Construction* (1964);

NSB 20, *Adobe/Puddled Earth* (out of print); *NSB 22, Stabilized Earth* (out of print);

Experimental Building Station, now National Building Technology Centre, Ryde, NSW.

Garrison, J and Ruffner, E (eds), *Adobe — Practical and Technical Aspects of Adobe Conservation*, Paper of the Heritage Foundation of Arizona, 1983.

George, Eugene, 'Adobe Bibliography', *APT Bulletin*, Vol 5, No 4, 1973, pp 97–103.

Hampshire County Council — Report on the Rebuilding of the Chalk Cob Boundary Wall to the Staff Car Park at Andover Cricklade College, C Stansfield Smith, County Architect, Winchester, February, 1984.

Harrison, J R, 'The Mud Wall in England at the Close of the Vernacular Era', *Transactions of the Ancient Monuments Society*, Vol 28, 1984, pp 152–74.

Harrison, J R, 'Traditional Cob and Chalk-Mud Building in the UK and Eire, A General Survey', unpublished dissertation, Diploma in Conservation Studies, Institute of Advanced Architectural Studies, York, 1979.

Hughes, Richard, 'Material and Structural Behaviour of Soil Constructed Walls', *Monumentum*, March 1983, pp 175–88.

McCann, John, *Clay and Cob Buildings*, Shire Album 105, Shire Publications Ltd, Aylesbury, England, 1983.

Middleton, G F, *Earth-Wall Construction*, Bulletin No 5, Department of Transport and Construction, Experimental Building Station (NSW), Australian Government Publishing Service, Canberra, 1982.

Northcliff, Stephen, *Down to Earth — An Introduction to Soils*, Leicestershire Museum Publications No 52, 1984.

Pearson, Gordon T, 'Chalk: Its Use as a Structural Building Material in the County of Hampshire', unpublished thesis, Architectural Association School of Architecture, Postgraduate Diploma in Building Conservation, April 1982.

Preservation of Historic Adobe Buildings, Preservation Brief No 5, Technical Preservation Services, Preservation Assistance Division, National Parks Service, US Department of the Interior, US Government Printing Office, 1978.

Torraca, Giorgio, 'Brick, Adobe, Stone and Architectural Ceramics: Deterioration Process and Conservation Practices', *Preservation and Conservation: Principles and Practices*, Proceedings of the North American International Regional Conference, Williamsburg, Virginia, 1972, The Preservation Press, 1976, pp 143–65.

Williams-Ellis, C, 'Cottage Building in Cob Pise, Chalk and Clay, *Country Life*, 1919, p 104.

Williams-Ellis, C and Eastwick-Field, J and E, 'Cottage Building in Cob Pise, Chalk and Clay', *Country Life*, 1947.

4.15 HISTORICAL GLASS

Burgoyne, Ian and Scobie, Rachel, *Two Thousand Years of Flat Glass Making*, Pilkington Bros plc, reprinted November 1983.

Dodsworth, Roger, *Glass and Glass Making*, Shire Album 83, Shire Publications, Aylesbury, England, 1982.

Harrison Caviness, Madeline, *Stained Glass Before 1540 — An Annotated Bibliography*, G K Hall & Co., Boston, Massachusetts.

Honeyborne, D B, 'The Conservation of Medieval Coloured Glass', Building Research Establishment Seminar, *The Conservation of Stone and Glass in Buildings*, Notes reference B453/76, BRE, Watford, England, 1976.

Lee, L, Seddon, G and Stephens F, *Stained Glass*, Mitchell Beazley, 1976.

Newton, R G, *The Deterioration and Conservation of Painted Glass: A Critical Bibliography*, published for the British Academy by Oxford University Press as Corpus Vitrearum Medii Aevi Great Britain, Occasional Papers II, 1982.

The Preservation of Historic Pigmented Structural Glass (Vitrolite and Carrara Glass), Preservation Brief No 12, US Department of the Interior, National Park Service, Rocky Mountain Regional Office, Cultural Resources Division, US Government Printing Office, February 1984.

Van den Bemden, Yvette and De Henau, P, *Les Vitraux Anciens — Note Technique Visant A L'Etablissement D'un Cahier Des Charges Type Pour La Restauration Des Vitraux Anciens Et De Valeur*, L'Institut Royal du Patrimoine Artistique et du Corpus Vitrearum de Belgique, 1987.

4.16 ORGANIZATIONS

The Ancient Monuments Society, 12 Edwards Square, London W8

APT (The Association for Preservation Technology) PO Box 2487, Ottawa, Ontario, Canada, KIP SW6

(The) Architectural Association, 36 Bedford Square, London WC1 3EG (List of theses for the Postgraduate Course in Building Conservation available from the library.)

ASCHB (Association for Studies in the Conservation of Historic Buildings) *ASCHB Transactions*: Editor: Stephen Marks, Hamilton's, Kilmersdon, Near Bath, Avon

Association of Bronze and Brass Founders, 136 Hagley Road, Birmingham, B16 9PN, Tel: (021) 454 414

BEC Publications, Federation House, 309 Coventry Road, Sheldon, Birmingham B26 3PL, Tel: (021) 742 5121

Brick Development Association, Woodside House, Winkfield, Windsor, Berks SL4 2DP, Tel: (0944) 88 5651

British Cast Iron Research Association (BCIRA) Alvechurch, Birmingham, B48 7QB, Tel: (0507) 66414

British Chemical Dampcourse Association (BDCA), 16A Whitechurch Road, Pangbourne, Reading, Berkshire RG8 7BP, Tel: (073 57) 3799

British Foundry Association, Ridge House, Smallbrook Queensway, Birmingham B5 4JP, Tel: (021) 643 3377

British Non-ferrous Metals Federation, Crest House, 7 Highfield Road, Edgbaston, Birmingham BL5 3ED, Tel: (021) 454 7766

British Standards Institution, 2 Park Street, London WIA 2BS, Tel: (01) 629 9000

BSI Regional Sales offices: 61 Green Street, London W1; 195 Pentonville Road, London N1; 3 York Street, Manchester 2.

British Wood Preserving Association (BWPA), Premier House, 150 Southampton Row, London WC1B 5AL, Tel: (01) 837 8217

(The) Building Bookshop, 26 Store Street, London WC1E 7BT, Tel: (01) 637 3151

(The) Building Conservation Trust, Apartment 39, Hampton Court Palace, East Molesey, Surrey, KT8 9BS, Tel: (01) 943 2277

Building Research Establishment – Building Research Station, Garston, Watford, Herts WD2 7JR. Tel: (0927) 374 040

Building Research Establishment – Princes Risborough Laboratory, Princes Risborough, Aylesbury, Bucks HP17 9PX, Tel: (084) 44 3101

Cathedrals Advisory Commission for England, 83 London Wall, London EC2M 5NA, Tel: (01) 638 0971/2

Civic Trust, 17 Carlton House Terrace, London SW1Y 5AW, Tel: (01) 930 0914

College of Masons, 42 Magdalen Road, Wandsworth, London SW18 3NP, Tel: (01) 874 8363

Conservation and Land Division, New Crown Building, Cathays Park, Cardiff, CF1 3NQ, Tel: (0222) 85511

Conservation Bureau, Scottish Development Agency, 102 Telford Road, Edinburgh EH4 2NP, Tel: (031) 343 1911/6

Copper Development Association, Orchard House, Mutton Lane, Potters Bar, Hertfordshire EN6 3AP, Tel: (0707) 50711

CoSIRA (Council for Small Industries in Rural Areas), Queens House, Fish Row, Salisbury, Wilts SP1 1EX, Tel: (0722) 6255

Council for the Care of Churches, 83 London Wall, London EC2M 5NA, Tel: (01) 638 0971/2

Crafts Council, The Conservation Section, 12 Waterloo Place, London SW1Y 4AU, Tel: (01) 930 4811

(The) Dry Stone Walling Association, The Old School, Pant Glas, Oswestry SY10 7HS, Tel: (0691) 654019

English Heritage (Historic Buildings and Monuments Commission for England), Fortress House, 23 Savile Row, London W1X 1AB, Tel: (01) 734 6010

Fire Prevention Information and Publications Centre, Aldermary House, Queen Street, London EC4N 1TJ

Friends of Terra Cotta, P O Box 421393, Main Post Office, San Francisco, CA 94142

(The) Georgian Group, 37 Spital Square, London E1 6DY, Tel: (01) 377 1722

Hampshire Buildings Preservation Trust Ltd, The Castle, Winchester SO23 8UE, Tel: (0962) 4411

Historic Buildings and Monuments Commission for England (HBMCE) *See English Heritage*

Historic Buildings Council for Scotland, 20 Brandon Street, Edinburgh EH3 5RA, Tel: (031) 556 8400

Historic Buildings Council for Wales, Brunel House, 2 Fitzalan Road, Cardiff CF2 1UY, Tel: (0222) 465511

HMSO, 49 High Holborn, London WC1V 6HB, Tel: (01) 928 6977

Institute of Advanced Architectural Studies, University of York, The King's Manor, York, YO1 2EP, Tel: (0904) 24919

Institute of Corrosion Science and Technology, Exeter House, 48 Holloway Head, Birmingham B1 1NQ, Tel: (021) 622 1912

(The) Institute of Metals, 1 Carlton House Terrace, London SW1Y 5DB, Tel: (01) 839 4071

International Centre for the Study of the Preservation and the Restoration of Cultural Property (ICCROM) 13 Via di S. Michele, 00153 Rome RM, Italy, Tel: Rome 58 09021, Telex 613144 ICCROM

Ironbridge Gorge Museum Trust, Ironbridge, Telford, Shropshire TF8 7AW, Tel: (09245) 3522

Lead Development Association, 34 Berkeley Square, London W1X 6AJ, Tel: (01) 499 8422

National Building Technology Centre (formerly (Commonwealth) Experimental Building Station) PO Box 30, Chatswood, NSW, 2067, Tel: Sydney 888 8888

National Corrosion Service, National Physical Laboratory, Teddington, Middlesex TW11 0LW, Tel: (01) 977 3222

National Trust, 36 Queen Anne's Gate, London SW1H 9AS, Tel: (01) 222 9251

Paint Research Association, Waldegrave Road, Teddington, Middlesex TW11 8LD, Tel: (01) 977 4427

RIBA Bookshop, 22 Portland Place, London W1N 4AD, Tel: (01) 637 8991

Royal Commission for Historical Monuments (RCHME), Fortress House, 23 Savile Row, London W1X 1AB, Tel: (01) 734 6010

Sacketts (Books for Craftsmen), 36 Salisbury Street, Blandford, Dorset, Tel: (0258) 53654

Scottish Development Department, Historic Buildings Branch, 25 Drumsheugh Gardens, Edinburgh EH3 7RN, Tel: (031) 226 3611

Shire Publications Ltd, Cromwell House, Church Street, Princes Risborough, Aylesbury, Bucks HP17 GAJ

SPAB (The Society for Protection of Ancient Buildings), 37 Spital Square, London E1 6DY, Tel: (01) 377 1644

Stone Federation, 82 New Cavendish Street, London W1N 8AD, Tel: (01) 580 5588

Stone Industries, Ealing Publications Ltd, Weir Bank, Bray, Maidenhead SL6 2ED, Tel: Maidenhead (0628) 23562

(The) Textile Conservation Centre, Apartment 22, Hampton Court Palace, East Molesey, Surrey KT8 9AU

Timber Research and Development Association (TRADA), Hughenden Valley, High Wycombe, Buckinghamshire HP14 4ND

Triangle Bookshop (Architectural Association), 36 Bedford Square, London WC1 3EG, Tel: (01) 631 1381

Victorian Society, 1 Priory Gardens, Bedford Park, London W4 1TT, Tel: (01) 994 1019

(The) Welding Institute, Abingdon Hall, Abingdon, Cambridge CB1 6AL, Tel: (0223) 891162

Zinc Development Association, 34 Berkeley Square, London W1X 6AJ, Tel: (01) 499 6636